Photocatalytic Dye Degradation Using Green Polymer-Based Nanostructures

Principles and applications in wastewater treatment

Online at: https://doi.org/10.1088/978-0-7503-5517-9

Photocatalytic Dye Degradation Using Green Polymer-Based Nanostructures

Principles and applications in wastewater treatment

Muhammad Ikram
Solar Cell Applications Research Lab, Department of Physics, Government College University, Lahore, Punjab, Pakistan

Muhammad Ahsaan Bari
Department of Physics, Government College University, Lahore, Punjab, Pakistan

Junaid Haider
Tianjin Institute of Industrial Biotechnology, Chinese Academy of Sciences, Tianjin, China

IOP Publishing, Bristol, UK

ISBN 978-0-7503-5517-9 (ebook)
ISBN 978-0-7503-5515-5 (print)
ISBN 978-0-7503-5518-6 (myPrint)
ISBN 978-0-7503-5516-2 (mobi)

DOI 10.1088/978-0-7503-5517-9

Version: 20230301

IOP ebooks

British Library Cataloguing-in-Publication Data: A catalogue record for this book is available from the British Library.

Published by IOP Publishing, wholly owned by The Institute of Physics, London

IOP Publishing, No.2 The Distillery, Glassfields, Avon Street, Bristol, BS2 0GR, UK

US Office: IOP Publishing, Inc., 190 North Independence Mall West, Suite 601, Philadelphia, PA 19106, USA

Dedicated to my parents for their encouragement and support throughout my studies, as well as for teaching me to be independent and responsible; my uncle Mukhtiar Ahmad (Late) for his unlimited support and love during my entire studies.

Contents

Preface

The rapid enhancement in the population results in an increased demand for potable water. The use of dyes in industries is increasing constantly and their effluents result in the contamination of groundwater. Reports suggest that around 10% of the dyes are released into the environment without being treated and contribute to the primary causes of water pollution. In addition, the increased utilization of chemical fertilizers and pesticides in agricultural practices has further contributed to the deterioration of the environment. Drinking chemical- containing water increases the risk of developing major health issues like hormone disruption, cancer, kidney and cardiovascular issues, abnormal brain function, among many others. Moreover, wastewater has a detrimental effect on the ability of aquatic organisms to reproduce. It impairs the regeneration and reproduction capacities of fish and plants. As a result of consuming polluted water, animals are also susceptible to a wide range of ailments.

To overcome the above-stated problem, scientists have developed a wide range of water purification techniques to meet the global potable water requirement. To understand these wastewater purification techniques this book has been designed to help students, young researchers, and scientists. Primary, secondary, and tertiary approaches are thoroughly described and analyzed, along with an estimate of their operational costs. An in-depth discussion is also made on the principal causes of water contamination and how it affects people, animals, and plants. The factors affecting the techniques for purifying water are highlighted. Also, a comprehensive explanation of photocatalysis, its homogeneous and heterogenous types, and factors influencing its effectiveness is covered. The mechanisms of photocatalysis are explained clearly. To improve the effectiveness of the photocatalysts, certain approaches are investigated, including doping and heterostructuring. In addition, the different dye kinds according to their charge are explored. We cover the chemical makeup of dyes, the environmental risks they pose, and their uses. Furthermore, a green polymer as well as its uses in the degradation of dyes are addressed. Doping green biopolymers into metal, metal oxides, and metal sulfides has recently improved photocatalysis and opened up new opportunities. Lastly, we discuss various issues and potential solutions that could help with the development of photocatalytic degradation of organic dyes.

Acknowledgement

All praise to 'ALLAH', the Almighty, most Gracious, the most Merciful, and the Sustainer of the worlds, who sent us 'MUHAMMAD (PBUH)' as a blessing for the whole Universe and the best teacher with the ultimate source of wisdom 'HOLY QURAN'.

First and foremost, I want to express my gratitude to ALLAH, the Almighty, for bestowing upon me the health to publish my book. It gives me great pleasure to express my deepest gratitude and profound sense of obligation to my role model, **Dr Muhammad Ikram**, Assistant Professor in the Department of Physics, for his able guidance, enthusiasm, competent advice, constructive criticism, little castigation, constant encouragement, valuable suggestions, and painstaking supervision throughout my studies, research work, and completion of this manuscript.

Lastly, I would like to express my gratitude to my **loving parents, sisters**, and my friend **Muhammad Mustajab** for the motivation to take up this program of studies, financial support and their hands in prayers for my success, and great patience and helpfulness throughout the fairly long period of training at this institution. Finally, as is customary, the errors that remain are mine alone.

Muhammad Ahsaan Bari

Author biographies

Muhammad Ikram

Dr Muhammad Ikram obtained his Master's degree (MPhil Physics) from BZU Multan, Pakistan, in 2010. He obtained his PhD degree in physics from the Department of Physics, Government College University (GCU) Lahore, Pakistan, through the Pakistan–US joint project between the Department of Physics, GCU Lahore, and the University of Delaware, USA, in 2015. He served as the Deputy Director of Manuscript Science at the Punjab Textbook Board (Pakistan). Later (2017–present), Ikram joined the Department of Physics, GC University Lahore as an Assistant Professor of Physics and Principal of the Solar Cell Applications Lab. Ikram received the Seal of Excellence Marie Skłodowska Curie Actions Individual Fellowship in 2017 and 2020. In 2021 Ikram was included in the 2% top scientists from Pakistan announced by Stanford University. His research interests involve the synthesis and characterization of inorganic semiconductor nanomaterials, 2D materials for water treatment optoelectronics, and electrocatalytic applications.

Muhammad Ahsaan Bari

Muhammad Ahsaan Bari obtained his BS (Physics) degree from Bahauddin Zakariya University (Punjab, Pakistan) in 2016 and his Master's degree in Material Physics from Government College University, Lahore, Pakistan, in 2022. Bari completed his MPhil research work in the Solar Cell Applications Lab at GCU Lahore (Punjab, Pakistan) and later worked as a Research Associate in the same laboratory in the catalytic and photocatalytic degradation field of organic dyes.

Junaid Haider

Dr Junaid Haider received his Master's degree in Food Science and Technology from Jiangnan University Wuxi, China in 2016. He received his PhD degree in Biochemistry and Molecular Biology from the Tianjin Institute of Industrial Biotechnology, Chinese Academy of Sciences (TIB-CAS) in 2022. Currently, he is a Postdoc Fellow at TIB-CAS. His research interests are focused on protein engineering, enzymatic depolymerization of recalcitrant polysaccharides, and the development of polysaccharide-based nanocomposites for dye degradation and antibacterial activity.

IOP Publishing

Muhammad Ikram, Muhammad Ahsaan Bari and Junaid Haider

Chapter 1

Conservation and wastewater remediation

Water is an essential component of life, and having access to water is necessary for the continued existence of all forms of life in the world. The current global situation is one in which there is a shortage of fresh water. The most significant contributors to water pollution come from dyeing processes, pharmaceutical waste, and pesticides. Together, these three sources account for around 70% of water pollution. According to several reports, approximately 15% of dyes are discharged directly into water without going through any sort of treatment. This contributes to the spread of numerous diseases that are transmitted by water, such as skin conditions, typhoid, cancer, and diarrhea, amongst many others. In this context, a number of approaches to reducing water pollution have been explored, together with their respective modes of operation and associated costs. The treatment of wastewater with nanotechnology has proven to be an effective method, and the field itself is quickly developing into one that holds great technological promise. It would appear that utilizing nanomaterials in the dye removal process is an approach that achieves the desired results successfully. This chapter provides an in-depth discussion of the factors that contribute to water contamination as well as the negative effects that these factors have on human health. In addition, a comprehensive overview of the various treatment methods is presented, which can assist any person in making an informed decision regarding water purification. Finally, factors that are known to have an impact on the treatment process, such as pH and the presence of heavy metals, have been addressed.

1.1 History

Most of the world's surface is water, which not only supports life but is also considered to be the most important natural resource. However, roughly 98% of water is seawater and, because of the salt content, it cannot be used for drinking

doi:10.1088/978-0-7503-5517-9ch1

purposes. Only around 2% of the water in the world is fresh, but about 1.5% is frozen solid in glaciers and ice caps at the poles. Another 0.36% is located beneath the Earth in subterranean wells and aquifers. Lakes and rivers hold only about 0.036% of the world's total water supply. In the vast majority (76%) of homes in 45 developing countries, women and children are responsible for collecting water on behalf of their families, according to a study conducted by the World Health Organization (WHO) and UNICEF monitoring program. This is time that is not spent attending school, working at a job that generates revenue, or taking care of members of one's family [1]. In addition, because of human and industrial activity, certain freshwater resources already in existence are becoming progressively contaminated and inaccessible. The increasing contamination of freshwater systems worldwide with countless different natural and industrial chemical contaminants is one of the biggest environmental problems facing humanity right now [2]. The rate at which the global population is growing is worrying and, at the same time, the rate of industrialization is accelerating, leading to an increased demand for water. As a result, water is becoming increasingly valuable in many nations, and in some regions of the globe it has become an essential commodity. According to a recently published article in *Nature*, 8% of the population faces major threats to the security of their water supply [3].

Owing to industrialization and other human activities, an increasing number of pollutants, such as dyes, medicines, heavy metals, pesticides, phenols, fluoride, detergents, and insecticides, are making their way into water sources. More specifically, previously unrecognized threats to public health associated with the release of hormones, personal care products, pharmaceuticals, and other chemical substances into the water supply are coming to light, leading to an increase in the need for stricter legislation [2]. There are still a high number of water pollutants that need to be dealt with. New types of chemical compounds are constantly being developed and introduced to the market as a result of rapid industrialization. These new substances will eventually enter aquatic systems. Additionally, harmful germs carried by water are widespread across the planet. These potentially dangerous microbes make their way into streams through untreated storm drains, septic tanks, sewage, runoff from farms, and from a variety of businesses, including tanning and meat packing companies.

Recently identified toxins in the water may negatively impact the ecology of aquatic ecosystems and human health. It is crucial for human health to have access to safe drinking water that is devoid of both potentially harmful substances and infectious viruses. In addition to this, pollution of the water supply is a major problem following a natural catastrophe. A wide variety of natural calamities, including tsunamis, earthquakes, storms, floods, and volcanoes, significantly impact water quality. If novel approaches to the delivery of clean water cannot be discovered in the coming decades, a lack of clean water may result in social and political unrest, water wars, and illnesses such as cholera. Because of the growing concern that people have regarding water, governments and organizations all around the world have begun enforcing stringent restrictions against the contamination of water sources. In the modern world, there is an ever-increasing interest in the investigation and creation of innovative procedures for wastewater treatment.

1.2 Water pollutants

Before beginning a conversation regarding water treatment and reclamation, it is important to have a firm grasp of the qualitative and quantitative aspects of the contaminants that can be found in water. The kinds of contaminants found in wastewater are determined by the activities carried out in industrial, agricultural, and municipal settings that release wastewater. The solvated, colloidal, or suspended forms of these water contaminants are what persist in the environment. The nature of the many sorts of contaminants that may be found in water can be classified as either physical, microbiological, or chemical.

Suspended inorganic and organic contaminants in water can result in the formation of physical pollution. Water takes on a new hue, flavor, and odor as a result of these types of impurities. The high temperature that results from industries and nuclear reactors releasing their cold water into bodies of water is one example of physical pollution. It results in a lower amount of dissolved oxygen, which is harmful to the organisms that live in water [4].

Microbial pollutants include important toxins such as parasites, viruses, and harmful bacteria. The different kinds of microbes that thrive in wastewater, including viruses, bacteria, fungi, plankton, algae, amoebas, and various kinds of worms, may be the cause of different illnesses. Such contaminants originate from the waste products of both humans and animals. When they mix with agricultural or sewage drainage water, they are carried into the water system and infect people with various diseases, including cholera. Because of this, it is necessary to utilize disinfectants such as chlorine to eliminate these impurities from the water supply [4].

In addition, the danger of radioactive pollution is increased whenever there is nuclear activity or an attempt to eliminate nuclear waste. The potential exists for radioactive materials to infiltrate into aquatic environments, where they are ingested by living things and passed on to people, having a range of effects on their genetic composition [4]. Radium is the most significant kind of radiation and can lead to bone cancer. Additionally, a shift in physiological state is produced when radioactive elements are present in the water [5].

Chemical pollutants are formed when there is an abundance of acids, organic compounds, fluorides, dissolved salts, pesticides, fertilizers, and metals in an environment. Most metals, including some of the more dangerous ones such as mercury, lead, barium, and cadmium (Cd), are soluble in water to varying degrees. Even non-toxic metals, such as iron, copper, calcium, salt, and magnesium, can cause various illnesses when they build up in the body. For instance, an excessive quantity of salt renders water undrinkable, raises the danger of developing kidney and heart illnesses, as well as causing plants to be poisoned. Similarly, the bulk of organic materials can dissolve in water. Either they are insoluble compounds, such as insecticides and detergents, or they are natural matter that can be broken down by the microorganisms in the water. The presence of nitrogen and phosphorous, which are the primary components of fertilizers, enables aquatic plants to experience accelerated rates of growth when these elements are present in the water. This might lead to the premature ageing of lakes, in which the bodies of water eventually transform into marshes or dry ground [4].

1.3 Causes of water pollution

Agricultural activities, natural factors, industrialization, and insufficiently sanitary sewer treatment systems are the four main sources of water contamination.

To begin, the most significant source of water contamination comes from industrial activity. Some examples of industries that contribute to water pollution are distilleries, tanneries, paper and pulp mills, textile mills, food processing plants, iron and steel mills, nuclear power plants, and others. Several dangerous substances, including volatile organic chemicals, poisonous solvents, and organic and inorganic chemicals, may be released into the environment throughout the production process. If these waste types are released into aquatic ecosystems without first receiving the required treatment, they will affect water quality [6]. The industrial sector is a substantial contributor of dangerous pollutants, including arsenic, cadmium, and chromium, which are all important contaminants released in wastewater [7]. The volume of wastewater generated by industrial operations has been steadily rising in tandem with the accelerating rate of urbanization [8]. In addition, the industrialization-related water contamination induced by foreign direct investment is also a significant factor. There is a positive correlation between industrial water contamination in less developed nations and foreign direct investment in such countries. The following are some ways in which industry could contribute to water pollution. The pH of water can be altered by alkaline or acidic substances. Corrosion of the pipelines will occur if the water is acidic, which will cause water pollution. Depending on the type of acid which the water has been contaminated with, this corrosion may pose various risks to human health. In addition, the pollution caused by alkali substances can result in the formation of salts such as bicarbonate, chlorides, hydroxides, and carbonates. The presence of calcium, magnesium, and carbon in the form of bicarbonate gives water its hardness. The saltiness of the Earth is directly proportional to the presence of chloride compounds. Lead, mercury, cadmium, and arsenic are the heavy metals that are most commonly found in a range environments. Figure 1.1 illustrates the main categories into which heavy metals can be placed as well as the sources of these metals. Mercury is a mineral, and mercury compounds can combine with chemicals found in water and soil. Mercury compound pollution can cause the central nervous system to malfunction along with causing sleep issues, emotional anguish, forgetfulness, gingivitis, as well as inflammation of the kidneys. Several different types of manufacturing make use of cadmium, including the production of batteries and plastics. The presence of cadmium in contaminated water causes a variety of ailments, including those affecting the kidneys, lungs, heart, and bones. Additionally, lead industries that manufacture batteries are among the most significant contributors to lead contamination. Blooming is another name for the phenomena that occurs when water becomes green due to the presence of nitrates and phosphates substances. They provide the appearance of a coating of green weeds on the surface of dams, seashores, lakes, and other bodies of water that are not moving. It is present on the water's surface, blocks the entry of oxygen and affects all aquatic life. Carbon, nitrogen, and phosphorus are the three components that make up green algae.

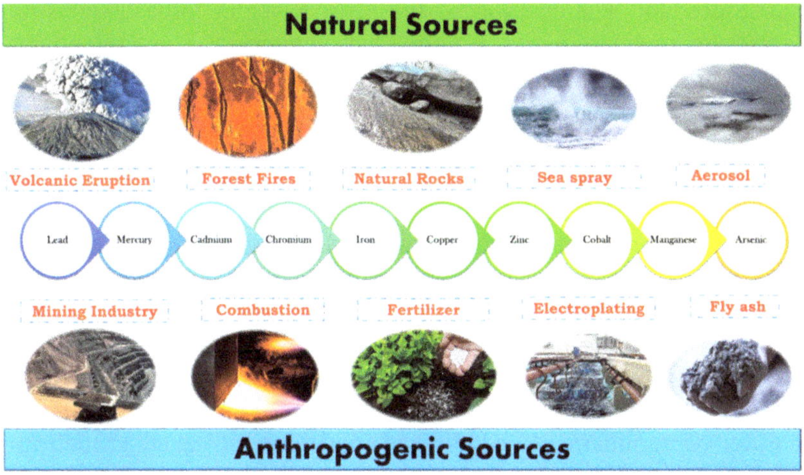

Figure 1.1. Broad classification and sources of heavy metals. (Reproduced with permission from [10]. Copyright 2021 Elsevier.)

It is important to remember that when nitrates bind with hemoglobin, they hinder the combination of oxygen, resulting in asphyxia. The elements magnesium and iron give water a rusty hue. Almost all bodies of surface water contain such algae. It is not harmful until it is found in extremely high amounts. Moreover, water pollution is caused by a wide variety of organic chemicals. The most well-known forms of pollution are caused by pesticides, fungicides, petroleum and its byproducts, and other industrial chemicals. Additionally, chlorine and fluorine are utilized in the water purification process to remove potentially dangerous bacteria. However, they have been shown to interact with hydrocarbons in the presence of cancer-causing chlorocarbons [9].

Second, a significant factor in the contamination of water supplies is agriculture. The use of nitrogen fertilizers, pesticides, and organic agricultural waste is mostly to blame for the significant water pollution that agriculture causes worldwide. The water used for agricultural purposes will get tainted with phosphate, nitrates, soil sediments, pesticides, pathogens, and salts due to these activities [11]. In addition, agriculture has caused irreparable harm to the natural condition of all of the world's freshwater systems [12]. However, in areas of underdeveloped nations such as India and China with limited access to water, untreated or inadequately treated wastewater is frequently used for irrigation. Both human health and the environment are at risk from this. A lack of adequate surface water has forced some developing nations such as China to employ wastewater irrigation long-term to meet their water needs for agricultural output. This has led to significant residues, pesticides, heavy metals, food, and agricultural land pollution, all of which represent a concern to human health and food safety [13]. The ingestion of pesticides through drinking water can negatively affect one's health. A 10% increase in pesticide use has been linked to a 1% increase in the medical disability index for people above 65, according to findings from the Health Life Expectancy Longitudinal Survey and information on pesticide use [14]. The example of the Musi River in India demonstrates that settlements who practice irrigation with wastewater have a greater prevalence of morbidity compared to homes that utilize regular water.

Third, natural forces are a contributor to the contamination of water. The water quality on the Chinese Loess Plateau shows a number of trace elements that are noticeably higher than the typical level found in other parts of the planet. The natural process of weathering, as well as man-made processes, can both be sources of trace elements. There is a potential for high salinity and salt concentrations in rivers with a low degree of good water quality [15]. The region of the Loess Plateau, situated at the center of China, is home to the highest incidence of one particular kind of water pollution known as hexavalent chromium contamination. This particular kind of pollution is brought into existence not only by the natural environment but also by the activities of humans. The primary contributors to surface water contamination are loess and mudstone; however, groundwater that contains high amounts of hexavalent chromium also plays a significant role in this issue [16].

Last but not least, the water supply and the sewage infrastructure are significant elements that influence potable water quality, particularly in underdeveloped nations. Along with China's rapid industrialization, economic growth, and urbanization, a lack of investment in basic water supply and treatment infrastructure has led to water pollution, an increase in the prevalence of parasitic and infectious diseases, and additional exposure to industrial chemicals, algal toxins, and heavy metals. Using an economic model, one can forecast the impact that water purification technology will have on the quality of water and, as a result, on human health. When water concentration in a household treated by water filtration technology falls from 100% to 90% of the overall amount utilized, up to 96% of the expected health benefits are lost. When the potential for contamination of the pre-treatment water supply is substantial, the decrease is much more noticeable [17]. To summarize, polluted water is the result of both natural processes and activities as well as the actions of humans. The water quality will be directly impacted by a wide range of human activities, such as climate change, population increase, industrial output, urbanization, and other elements [18], as well as religious activities [19]. One of the factors contributing to deteriorating water quality is unsuitable solid waste disposal, as well as gravel and sand [20].

1.4 Water pollution and the living world

There is a stronger connection between polluted water and health issues than previously thought. Microorganisms that may cause disease are referred to as pathogens, and these pathogens are directly responsible for the transmission of disease from one human to another. Some infections are prevalent worldwide, whereas others are confined to a particular location [21]. A number of waterborne diseases are transmitted from person to person [22]. Extreme weather conditions, such as flooding and heavy rainfall, contribute to the spread of various illnesses throughout developing and industrialized nations [23]. Globally, 10% of the population relies on vegetables and fruits grown using polluted water. There is a relationship between the fecal–oral route of infection and fecal contamination of water sources, which is responsible for several aquatic infectious illnesses. There are a variety of illnesses that have been linked to the use of dirty water, including those that affect the respiratory system, the digestive tract, the nervous system, and the cardiovascular system. Nitrogenous chemicals are the underlying factor in cancer development and blue baby syndrome [24]. The death rate from cancer is substantially higher in rural areas than in urban areas since urban residents drink water that has been treated, however, people living in rural areas do not have access to potable water and drink water that has not been processed. Because of problems with sanitation, hygiene, and access to clean water, persons living in poverty have a significantly increased chance of contracting a disease [25]. It has been shown that pregnant women exposed to chemicals throughout their pregnancy are more likely to have babies with a low birth weight [26], which in turn has a severe impact on the fetus's health. Contaminated water also has a wide range of other detrimental impacts. The production of crops is harmed by water of insufficient quality, which can lead to the contamination of our food, which poses a threat to both human and aquatic life [27]. Contaminants cause disruptions in the food chain, and heavy metals, in particular iron, have a negative impact on the respiratory systems of fish [22]. When humans consume fish that have iron clogging in their gills, which is fatal to fish, it produces severe health problems [28]. Loss of hair, liver cirrhosis, renal failure, and brain dysfunction are all caused by drinking water polluted with metals [29]. The broad categories of water contaminants are shown in figure 1.2, along with their impact. These potentially dangerous contaminations have been divided into two groups: chemical and microbial contamination.

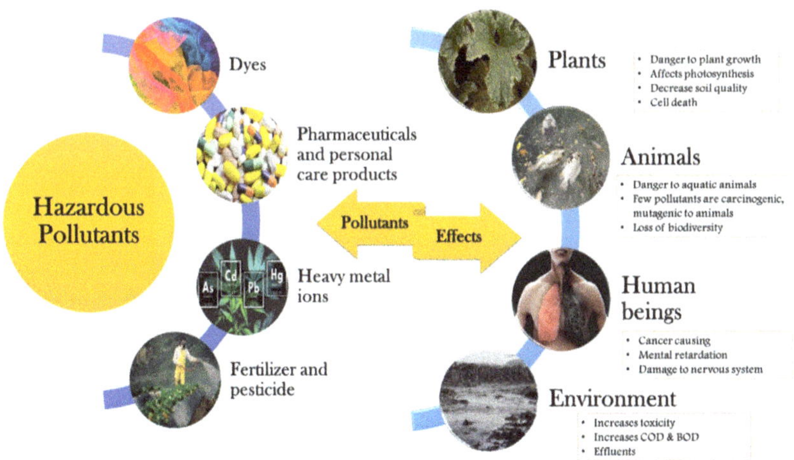

Figure 1.2. Sources and causes of water pollution. (Reproduced with permission from [10]. Copyright 2021 Elsevier.)

1.4.1 Chemical contamination

In most cases, the numerous types of chemical compounds that are discharged into the environment by factories, workshops, irrigated areas, tanneries, and residential wastewater are to blame for various ailments. Volatile chemicals, heavy metals, organic materials, hydrocarbons, pesticides, and other pollutants can all fall under this category. The contamination may induce infectious disorders such as skin cancer and chronic dermatoses, as well as lung infection and eye discomfort if exposed to enough. Most of them cannot be broken down by natural processes or absorbed by living organisms. Because of this, they are able to remain in the portable water for a long time and have the potential to readily accumulate in our system of food chains. Nonylphenol, estrone, estradiol, and ethinylestradiol are potentially harmful chemical compounds that might be present in many surfactants and pharmaceutical personal care products that are currently on the market. These compounds are endocrine-disrupting substances, as well as very modest concentrations of these substances in humans can have serious negative effects on the body [30]. The presence of perfluorinated compounds (PFCs), which are harmful in their natural state, has been documented in large amounts worldwide in wastewater [31]. Additionally, exposure to PFCs might cause major health issues, including pre-eclampsia, birth abnormalities, decreased human fertility, immunotoxicity [32], neurotoxicity [33], and carcinogenesis [34].

1.4.2 Microbial contamination

Researchers have shown that exposure to the microbiological pollutants present in wastewater that has not been treated poses substantial health hazards. Campylobacteria, typhoid fever, diarrhea, encephalitis, hepatitis A, giardiasis, gastroenteritis, salmonellosis, and poliomyelitis are only a few serious health problems that may be caused by the wide variety of germs [35]. *Salmonella typhimurium*, *Escherichia coli*, *Pseudomonas aeruginosa*, *Shigella sonnei*, *Giardia intestinalis*, *Legionella* spp., and *Vibrio cholerae* are only a few bacterial species that have been linked to the transmission of waterborne infections and acute sickness in humans [36]. These above-listed microbes have the potential to escape into the environment through animal husbandry, public sewage water networks, or hospitals, and then make their way into the food system through water distribution systems.

1.4.2.1 Diarrhea

Diarrhea is the most common symptom of diseases affecting the digestive tract and the most common ailment that may be traced back to contaminated water. Diarrhea is one of the common leading causes of illness and death in young children, particularly in countries where the average income per person is low. In underdeveloped countries, diarrheal diseases cause 21% of deaths yearly in children under 5 years old [37]. A significant number of the infectious agents that might cause diarrhea are directly connected to the presence of contamination in the water [38]. When people drink water that has not been purified, they put themselves at risk of contracting illnesses that are caused by parasitic worms [39]. It was shown that drinking treated from water filtration systems is related to a decreased incidence of diarrhea in people of all ages compared to drinking untreated water [40]. A study conducted in Brazil (southern region), for example, found several factors significantly associated with an increased risk of mortality from diarrhea, such as a shortage of flush toilets and plumbed water, overcrowded households, and poor housing conditions. Compared to families with access to piped water, those without it had a 4.8 times higher risk of neonatal mortality from diarrhea.

There are examples of enteroviruses found in aquatic environments. More than one hundred dangerous viruses are released into the environment when humans and animals defecate. These viruses are spread throughout the ecosystem via sewage treatment plants, estuarine water, rivers, groundwater, seawater, drinking water, private wells, and water that has not been adequately treated. According to research carried out in Pakistan, coliform contamination was discovered in a few of the country's water sources. The primary contributors to the contamination of drinking water are the improper disposal of solid waste and sewage, the excessive use of fertilizers and pesticides, and the deterioration of pipelines. The presence of coliform bacteria in polluted water is the most significant cause of water-related illnesses in this region, including viral hepatitis, dysentery, gastroenteritis, and diarrhea [41]. As a result, preventing the transfer of diarrheal infections from the environment to humans should be seen as an essential function of health interventions related to sanitation and water [37].

In research on water quality and diarrhea, the approach utilized most frequently is called a meta-analysis. It was found that improving sanitation and expanding access to safe water might reduce diarrheal disease prevalence overall by 26%. In Malaysia, having access to fresh water and sanitation has been linked to an 82% lower incidence of neonatal deaths. This was especially true for infants who had not been nursed before death. All of the treatments that improved sanitation and water quality greatly decreased the danger of diarrheal illness, and it has been demonstrated that water quality interventions were far more successful than previously thought. There was no significant difference in effectiveness between single-focus and multiple treatments (which included steps to improve water and sanitation systems) [42]. Interventions aimed at improving water quality lowered the prevalence of diarrhea in children and decreased the likelihood of *E. coli* contamination in preserved water [43]. It has been shown that measures taken to enhance water quality are typically successful in avoiding diarrhea in children of all ages, including those less than 5 years old. Nevertheless, the results of several studies demonstrated a substantial degree of variability, which may be attributable to the research procedures and the settings of the trials [40].

1.4.2.2 Skin diseases

Research conducted as early as the 1950s shows that swimming is not necessarily good for health. In fact, the overall infection rate in the swimming group was substantially higher than that in the non-swimming group. According to the survey results, the prevalence of illness is about 100% greater in persons under the age of ten than in those older than ten. There is a specific percentage that is accounted for by skin illnesses. During the summers of 1986 and 1987 in Hong Kong, a prospective epidemiological investigation on beach water contamination was carried out. According to the study's findings, swimmers on the coastline beaches of Hong Kong are more likely than non-swimmers to complain of systemic diseases, including of the eyes and skin. The danger of developing skin disorders and other ailments is significantly increased when one swims in beach waters with a greater pollution level. There was a correlation between the cleanliness of the beach and the incidence of swimming-related illness symptoms.

According to research conducted in arsenic-affected communities in the Sindh region of Pakistan's south, skin problems are brought on by poor water quality. Compared to persons who drank urban water with low levels of arsenic, people who drank high levels of arsenic drinking water had much higher levels of arsenic in their hair. This fact was uncovered by researchers looking into the link between high levels of arsenic in freshwater, which is caused by pollution in water sources, and skin problems (primarily keratosis and melanosis). The amount of arsenic found in drinkable water has a direct and detrimental influence on the life of local communities, and the most common clinical manifestation of arsenic poisoning is a skin disease. There is a relationship between arsenic-tainted drinking water and the quantities of arsenic found in biological samples (blood and hair) taken from people who suffer from skin illnesses [44]. Another study conducted in Bangladesh discovered that the contamination of rivers causes scabies in a significant number of residents [45]. In addition, the contamination of water sources by industry can potentially cause skin cancer [46].

The use of meta-analysis in research has revealed that prolonged exposure to contaminated recreational waters in marine environments can have negative effects, including recurrent skin pain (such as rashes or itching). Many different types of infectious bacteria might be the culprits behind skin problems in swimmers [47]. People (non-swimmers and swimmers alike) who were exposed to water with bacteria concentrations over the threshold level had a greater relative chance of acquiring skin illness, and the levels of bacteria in saltwater have been shown to have a strong correlation with skin symptoms.

According to the findings of several studies, those who swim frequently are also more prone to suffer from skin problems than those who do not. This discrepancy could be attributed to swimmers' 'risk perception bias', since they are more likely to notice and report skin conditions and are generally more aware that such exposure could have negative health implications. Additionally, it is possible that swimmers exaggerated their symptoms and claimed disorders that others would not classify as true skin conditions [48].

1.4.2.3 Cancer

Statistics compiled by the WHO indicate that the number of people diagnosed with cancer in the year 2020 reached 19.3 million, while the number of people who passed away as a result of cancer rose to 10 million. Currently, one in five individuals who experience a fever may risk developing cancer. The pollution of the water source, the techniques used to treat the water, and when the water is provided to consumers are the three primary entry points for carcinogens into drinking water, each of which can result in a different mix of carcinogens and varying concentrations of those carcinogens.

It has been demonstrated that certain toxins found in water, such as arsenic, nitrate, chromium, and others, have substantial ties to elevated cancer risk. Arsenic in drinkable water poses a risk of developing bladder cancer, kidney cancer, and skin cancer in humans [49]. The cancer risk in the general public posed by arsenic prevalent in the United States water system may be comparable to the risk posed by secondhand smoke and radon found in private homes. However, a person's likelihood of being affected by arsenic's cancer-causing properties varies greatly from person to person. A controlled experiment conducted in the north of Chile between the years of 1994 and 1996 involved patients diagnosed with lung cancer and a hospital with a frequency-matched patient population. The experiment's findings demonstrated a strong connection between lung cancer and the presence of arsenic in potable water. However, when arsenic exposure levels were lower than 0.64 mg l^{-1}, this effect was minimal [50]. In addition, research has shown that the combined effects of smoking and ingesting arsenic from drinking water are far more likely to result in lung cancer [51].

Nitrates are a more widespread pollutant that has been found to have a stronger connection to human malignancies, particularly colorectal cancer. An investigation conducted in East Azerbaijan concluded that a strong link exists between nitrate and colorectal cancer in males. There is a correlation between the concentration of nitrates and their potential to cause cancer. The danger is greatly increased when the levels of lead in drinking water reach 3.87 mg l^{-1}, which is much below the existing threshold for drinking water, which is 50 mg l^{-1}. Consuming water with negligible nitrate contents compared to the existing drinking water guidelines is another factor contributing to an increased risk of colorectal cancer [52]. In addition, studies have shown that hexavalent chromium does have the capability of causing lung cancer in people when it is ingested in the form of drinking water [53]. A case study conducted in Taiwan (Changhua County) also showed that excessive amounts of chromium in the environment are connected with stomach cancer [54]. There is a correlation between the number of people who pass away from cancer and the quantity of trihalomethane (THM) found in the water used for drinking. There was a significant correlation between THM levels and brain cancer mortality, bladder cancer mortality, kidney cancer mortality, and non-lymphoma Hodgkin's mortality in males. The death rate from bladder cancer was found to have the most significant and consistent correlation with the THM exposure index. Carcinogens may be released during the chlorine treatment process, which is one of the steps in the water treatment process. Contaminated drinking water is related to all types of cancer, including malignancies of the urinary tract and the digestive system.

Herbicides, fertilizers, and pesticides, which release and contain nitrates, are examples of the many additional types of pollutants that may be found in water that are also thought to cause cancer [55]. In Hebei, China, it was discovered that the usage of nitrogen fertilizers in agriculture was directly associated with the contamination of three nitrogen compounds in well water, and this was highly correlated with mortality from esophageal cancer [56]. The impact of watershed pollution on the onset of cancer is also quite location-dependent due to the time-lag effect. The rate of esophageal cancer mortality brought on by polluted water is significantly higher downstream than in other areas of the country due to the cumulative effects of decades' worth of pollution [57]. A grade 6 drop in water quality was linked to a 9.3% increase in the frequency of fatalities from intestinal cancer, according to the results of a study that looked at differences in the quality of water throughout the watersheds [58].

1.5 Techniques for water treatment

Treating wastewater, combining physical, biological, and chemical processes, removes contaminants from untreated wastewater. Numerous systems incorporate several wastewater treatment methods stages, classified as tertiary, secondary, and primary wastewater treatment. This is done in order to accomplish varying degrees of contaminant removal. The early steps of the filtration technique, which could be either chemical or physical, are referred to as 'primary treatment'. Secondary treatment involves biological wastewater treatment. In the tertiary treatment methods polluted water that has been treated in the physical and biological treatment processes is transformed into high-quality water that may be utilized for a variety of purposes, including the supply of drinking water, industrial water, medicinal water, and so on. Up to 99% of the contaminants are removed from the water during the tertiary process, and the water is then changed into a quality that is suitable for a particular use. These three procedures are often linked to one another in a comprehensive water treatment facility to produce water of high quality and without risk. This section provides a review of the various traditional as well as cutting-edge solutions which are currently in operation and also an explanation of how these technologies are utilized for the efficient treatment and purification of wastewater.

1.5.1 Primary/physical water treatment technologies

Methods such as filtration, screening, sedimentation, centrifugation, coagulation, flotation, and gravity are utilized to carry out the primary stage of water treatment in this category. In most cases, these procedures are employed when the water has a very high level of pollution. The following subsections provide condensed explanations of these various approaches.

1.5.1.1 Screening, filtration, and centrifugal separation

Wastewater contains a significant quantity of particles and grit, both of which have the potential to cause harm to the treatment apparatus. This is something that the screening procedure can address. A screen is a device with apertures, which are typical of the same size, used to separate coarse materials suspended in wastewater and solids floating in the water. The screening element is often made up of parallel wires, rods, or bars, and the apertures are typically either rectangular or circular slots. The narrower the opening of the screen, the higher the number of screenings that may be performed [59]. Grit is made up of substantial inorganic substances that weigh more than organic biodegradable particles in wastewater. Grit includes things such as eggshell, metal fragments, and sand. Eliminating grit is important to minimize unnecessary wear and abrasion on mechanical equipment. A grit chamber can have either a vertical or horizontal flow, and its cleaning can be done either manually or automatically. The grit that is produced in a chamber that has been constructed and is being operated correctly is free of organic materials and can be utilized as landfill [60]. If the grit in question contains a significant amount of organic material, it will either be buried or converted into manure before being disposed of.

In the process of filtration, water is forced through a medium that is characterized by the presence of minute holes. For this purpose, a set-up that has pores measuring between 0.1 and 0.5 μm in size on average is typically utilized. The elimination of suspended particles, greases, oils, microorganisms, and other impurities is one of its many applications. Several kinds of filters may be employed, including cartridges and membranes. The filtration procedure may be utilized to eliminate particles with a size of less than 100 mg l^{-1} and oil with a size of less than 25 mg l^{-1}, both of which can result in a reduction of as much as 99%. In order to purify water, a filtering procedure is often used. The processes of ion exchange, adsorption, and membrane separation all use water generated through filtering. Additionally, filtering systems are responsible for producing drinkable water [61].

The removal of suspended noncolloidal materials requires the use of centrifugal separation (size up to 1 μm). The polluted water is pumped through centrifugal machines, which are then rotated at varying speeds to separate the liquid portion from the particles (sludges) of the wastewater. The relative densities of suspended particles are directly proportional to the degree to which they are separated. In addition, the removal of suspended materials is also attributable to the speed at which the centrifugal machine operates. There are a variety of applications, some of which include the separation of grease and oils and source reduction. The counter-flow machine, solid-bowl machine, basket-type machine, and counter-current flow machine are the many kinds of centrifugal machines that may be purchased and put into operation [62].

1.5.1.2 Sedimentation and gravity separation

Sedimentation is used to remove suspended particles from wastewater. These particles include flocs, sand, and clay. This process occurs on its own whenever the water is not moving because gravity causes the larger particles to sink to the bottom and form a sludge layer. The time it takes for solids to settle is determined either by the size and density of the particles or, if the water is moving, the velocity of the water. The addition of alums often helps the process of sedimentation along. Only a 6% reduction in the suspended solids is achievable by the use of gravity separation. Sedimentation decreases the need for flocculation and coagulation, which is one of its advantages. Additionally, sedimentation can be done following coagulation to improve the effectiveness of the continual filtering involved in the process [63]. Tanks that are specifically designed for sedimentation are required for wastewater treatment. A sedimentation tank offers the particles the necessary support they need to guarantee that they sink to the bottom of the tank. Several types of sedimentation tanks can be used, including multi-layer, horizontal flow, radial flow, settler, and floc blanket sedimentation tanks.

1.5.1.3 Coagulation/flocculation

The removal of heavy metals by the physical processes of flocculation and coagulation is very effective. During this process, tiny colloids and particles agglomerate into bigger ones, reducing the turbidity, natural organic matter (NOM), and other contaminants found in wastewater. The first thing that needs to be done is to mix a coagulant into the water. This causes the colloidal particles in the water to clump together to form flocs, which are very small aggregates [64]. Polymeric ferric sulfate (PFS), aluminum sulfate, ferric sulfate, polyacrylamide (PAM), and polyaluminum chloride (PACl) are some of the coagulants that are utilized commonly [65–67]. In the second step, the flocs will agglomerate while being subjected to mild agitation. After this, they will settle and then be disposed of as sludge. Polyelectrolytes (PE) have reportedly been one of the most useful flocculants, although the sludge produced may be harmful [68]. The flocculants are often man-made and do not break down in natural environments [69]. Figure 1.3 depicts the process, including the sedimentation that occurs. Specific drawbacks include the health hazards of inorganic coagulants and toxicity, a high amount of sludge, selection for some metals but ineffectiveness for emerging pollutants, rising effluent color, inefficiency when utilizing natural coagulants, and the complexity of scaling up [64]. This process is often used to remove heavy metals such as Ni^{2+}, Pb^{2+}, and Cu^{2+}. Additional metals, such as Se^{2+}, As^{2+}, Sb^{3+}, Cr^{2+}, Ag^{2+}, and Sb^{5+}, can also be removed effectively. Due to the adaptability of this procedure, it may be employed either as a post-treatment, a pre-treatment, or the main wastewater treatment [64].

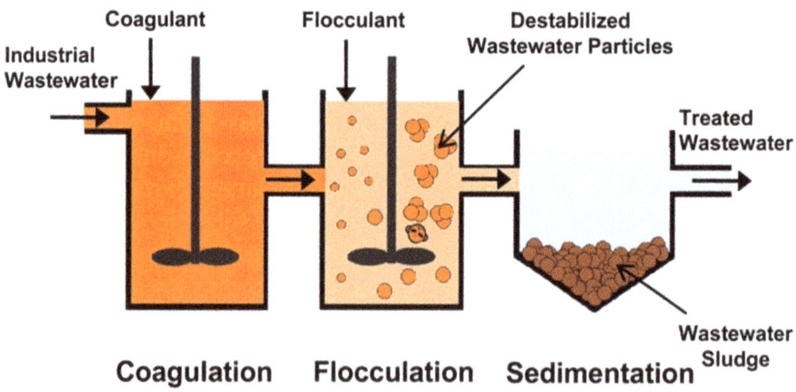

Figure 1.3. A schematic of the coagulation–flocculation treatment process. (Adapted with permission from [70]. Copyright 2022 American Chemical Society.)

1.5.1.4 Flotation

The use of the flotation process can eliminate several different types of metal ions. Figure 1.4 provides an overview of the flotation process's basic flowchart. A significant amount of research was put into ion flotation, precipitation flotation, and dissolved air flotation. In the process known as dissolved air flotation, gas (or air) is introduced into polluted water to produce microbubbles. These microbubbles can affix themselves to a metal ion, forming agglomerates with a lower density and the subsequent raising of flocs in the wastewater. The slug accumulated at the top surface is not difficult to remove [71]. The ion flotation procedure depends on the elimination of hydrophobic species by air bubbles because surfactants are used, which increase the hydrophobicity of metal species. The additional surfactants serve as collectors, while others are responsible for controlling the ion flotation indices [72]. When there is a small amount of metal ions in a large volume of polluted water, ion flotation appears less effective [73]. The ion flotation method demonstrates low energy usage levels, decreased sludge quantities, selective treatment, and restricted volume demands [74]. Precipitation flotation is a chemical precipitation process that employs microbubbles in its most basic form. Precipitation flotation can be effectively completed in a short period of time [75]. The flotation processes, in general, provide benefits, such as speedy operation, a compact process, and a relatively modest cost. Because surfactants serve as collectors in ion flotation, it is necessary to use surfactants that are both effective and non-toxic. Chemically synthesized surfactants have been developed because of

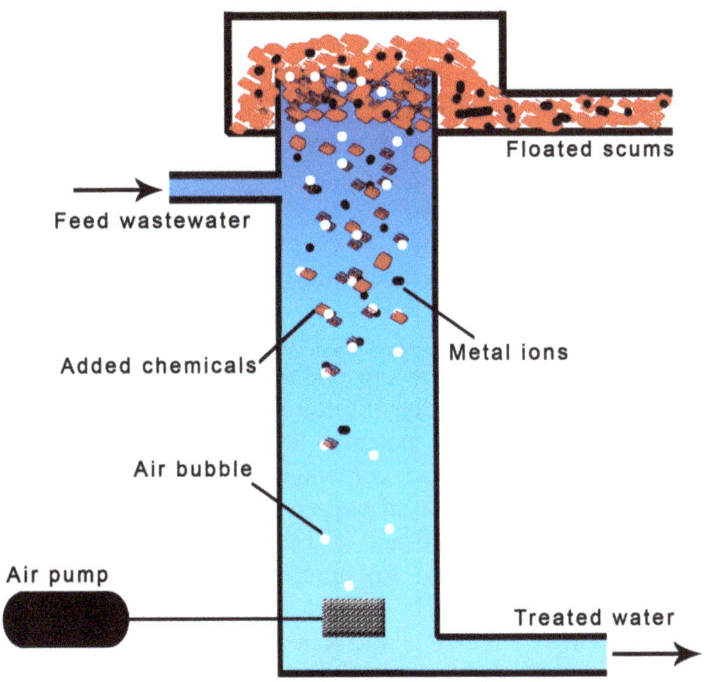

Figure 1.4. An illustrative schematic of the flotation treatment process. (Reproduced from [76]. Creative Commons Attribution 4.0 International License http://creativecommons.org/licenses/by/4.0/.)

their powerful collecting ability, excellent selectivity, and straightforward manufacture. The problems of expense and toxicity, however, prevent their widespread use. On the other hand, biosurfactants could seem like a better option for the environment, but studies have shown that they have low removal rates, huge dose quantities, and lengthy processes. In this context, nanoparticles have been suggested as potential novel collectors due to their ability to exhibit advantages similar to synthetic and biosurfactants. Electro-flotation, also known as electro-flocculation, is a method that has proven beneficial for recycling and treating wastewater. This technique can remove up to 99% of oil and grease and up to 75% of suspended particles. It is a productive method for the cleanup of wastewater from establishments such as paper mills as well as refineries.

1.5.2 Secondary/biological water treatment technologies

In the second stage of treatment, microorganisms are used to break down the organic content in the water and produce inorganic byproducts. The microbes are then removed using the sedimentation method. The digestion of wastewater, both aerobically and anaerobically, is a part of the biological treatment. First, there is the activated growth process, also known as the fixed film process, which includes an intermittent filter, rotating biological contactor, trickling bed, and contact bed. This is a method for converting organic matter that involves the biomass becoming attached to an inert medium, such as ceramic, rock, slag, or plastic [77]. Second, suspended growth procedures include aerated lagoons, activated oxidation ponds, and sludge processes. These processes include suspending the microbial cells either mechanically or naturally to convert organic waste [78]. Because of this, the leftover suspended and organic particles from the physical method are eliminated during the biological procedure, making the water consistent enough for chemical treatment.

1.5.2.1 Aerobic processes

The process of aerobic decomposition, which is produced by aerobic and facultative bacteria, takes place in biodegradable organic matter when air or oxygen is readily present in the polluted water in the form of dissolved gas [79]. The magnitude of the process is dependent on the amount of oxygen that is present, the length of time that the substance is retained, the temperature, and the biological activities carried out by the bacteria. In addition, the inclusion of certain chemicals that are necessary for the development of bacteria has the potential to hasten the process of biological oxidation, which removes organic contaminants from the environment. The approach efficiently removes chemical oxygen demand (COD), biological oxygen demand (BOD), nitrates, dissolved and suspended organics, phosphates, volatile organics, and other contaminants. An up to 90% reduction in the amount of biodegradable organics present can be accomplished through the use of this strategy. Generating a substantial number of bio-solids, which calls for additional, more expensive treatment and management, is one of the drawbacks of using this technology. Trickling filters, activated sludge processes, and oxidation ponds are the three methods that are used to complete the aerobic process.

1.5.2.1.1 Activated sludge

After primary sedimentation, a biological process known as 'activated sludge' occurs. After the initial treatment, some colloidal and suspended particles are still present in the collected wastewater, and these particles are stirred up by the presence of air. The floating solids eventually grow into larger solids, whereas the colloidal solids serve as the foundation upon which biological life may grow [80]. Activated sludge is typically a brownish substance that resembles foc (fraction of organic carbon in primary solids) and is made up of organic matter that is derived from wastewater. This substance is also home to many bacteria and other types of biological life. The live organisms that make up activated sludge can absorb or adsorb both colloidal and dissolved organic materials. This is one of the many beneficial properties of activated sludge. Living organisms utilize the absorbed substance as food and convert it into inert, insoluble solids, and new bacterial cells. The majority of this transformation is carried out in stages. Certain bacteria consume the original complex molecules to generate simpler chemicals as waste products [81]. Further bacteria then use the resulting waste material to produce even simpler chemicals. This process is repeated until the bacteria can no longer depend on the ultimate waste products. Some sort of agitation must be utilized to keep the activated sludge from sinking to the bottom of the tank as it interacts with the contaminated water being treated. This agitation will maintain the suspension of the activated sludge.

1.5.2.1.2 Trickling filter

The use of a trickling filter is an example of an aerobic treatment technique. With this process, the microbes used for treatment connect to a medium in order to remove organic debris from contaminated water. Rock, sand, redwood, and even a variety of man-made materials can all have microorganisms clinging to them. The microbes that develop on the filtering media are referred to as 'biological slime', and they include bacteria, fungus, and algae. As part of the process, the contaminated water is sprinkled and afterwards allowed to trickle through the medium. During this time the biological slime that is adhered to the medium will metabolize the organic material that is already present in the wastewater. The biological slime's thickness increases due to the conversion of organic material contained in contaminated water towards newer cellular material. After some time, the biological slime layer will get quite thick and eventually separate from the surface. Sloughing is the name given to the process that describes the separation of the slime layer. To separate the particles from the liquids, this detached slime layer is recovered by under drainage together with the treated wastewater [82].

1.5.2.2 Anaerobic processes

The process of anaerobic decomposition, also known as putrefaction, occurs when there is insufficient free dissolved oxygen in the wastewater. The complex organic substance is broken down by anaerobic and facultative bacteria into simpler organic molecules based on sulfur, carbon, and nitrogen. Hydrogen sulfide, ammonia, methane, and nitrogen are four significant gases produced due to this process. This technique is utilized in order to lessen the biological burden that wastewater carries [83–88].

1.5.3 Tertiary/chemical treatment

It is necessary to subject wastewater to tertiary treatment in order to get rid of the contaminants, nutrients, and pathogens that have persisted past the primary treatment, to conform to quality standards for water reuse, and to make the treated wastewater suitable for direct release into bodies of water as well as land applications. Some examples of water bodies include rivers, lakes, and ponds. Electrodialysis, crystallization, evaporation, distillation, solvent extraction, coagulation, oxidation, electrolysis, reverse osmosis, adsorption, ion exchange, and precipitation are the processes that are utilized in order to accomplish this goal. This approach also includes advanced oxidation, which tries to modify the compound's specific harmful effect by influencing the components required for the hazardous action. These procedures are broken down in more detail below.

1.5.3.1 Distillation

During the distillation process, the water is heated until it reaches boiling point, which is 100 °C. At this temperature, the contaminants in the water evaporate, leaving the cleansed water behind [89]. The resulting vapors are condensed into a liquid state by cooling. The wastewater must be free of flammable pollutants, and the water generated using this method is almost completely free of contaminants. This method uses several different kinds of boilers, including those with multistage and double distillation. The quantity of water o be treated determines the size of the boilers that are needed. In water purification and rehabilitation, distillation applications include the provision of water for use in labs and the manufacture of medicines. Additionally, distillation is a practical technique for producing drinking water from brackish and salty water, both of which are prevalent in nature.

1.5.3.2 Crystallization

In the process of crystallization, pollutants are eliminated from the environment by increasing the concentrations of the pollutants to a level at which the pollutants begin to crystallize. This condition can arise due to evaporation, a decrease in the water's temperature, or adding additional solvents to the mixture. This approach can suit the treatment of wastewater containing high total dissolved solids (TDS), such as soluble inorganics and organics. Through this process, many additional substances, including sulfite, ammonia, and bicarbonate, decompose into different gases. As a result, crystallization is sometimes employed for pH regulation. Crystallization is a common treatment method for wastewater produced by industries such as paper manufacturing and dying, as well as cooling towers that utilize coal or gas for fuel. In addition, it may be utilized for source reduction. For the process of crystallization, methods such as draught tube baffles, forced circulation, fluidized suspensions, and surface-cooled crystallizers are utilized. The wastewater treated with this procedure comes out quite clean and fresh [89].

1.5.3.3 Evaporation

Evaporation is a naturally occurring phenomenon that, in most cases, is used to minimize the volume of waste liquid, but in more recent times it has also been utilized as a means of water purification [89]. Under normal circumstances, the molecules that make up the water's surface will evaporate, and the water that results from the condensation of these evaporated molecules will be completely pure. For the process of recycling water, mechanical evaporators are also utilized. Vacuum evaporation has occasionally been used in the wastewater recycling process. With the exception of volatile organic contaminants, evaporation is an excellent approach for removing both organic and inorganic pollutants, and it retains its efficacy even at exceptionally high levels of pollutants (nearly 10%). Scaling, fouling, and foaming, in addition to the presence of carbonates and suspended particles, are the primary issues with utilizing this method because they all contribute to maintenance issues. Evaporation applications include the purification of wastewater containing petroleum and fertilizer, as well as wastewater from the food processing and pharmaceutical manufacturing sectors. Additionally, it is utilized to deliver water to membrane processes and ion exchangers. After evaporation, the remaining water is used to cool boilers and towers. It can be used as a technique of water source reduction.

1.5.3.4 Solvent extraction

Solvent extraction involves adding organic solvents to wastewater to remove contaminants; these solvents are incompatible with water and can dissolve pollutants. Benzene, hexane, and acetone, as well as many other hydrocarbons, are among the most frequently utilized solvents. Sometimes, a tiny portion of the solvent will remain combined with the water, and in these cases the distillation method can be used to recover the solvent. This method only removes organics, oils, and grease from the surface. However, this approach has the potential to remove specific metal ions as well as actinide compounds. Recycling and water purification are two of the applications. Additionally, it has been utilized in water source reduction applications [89].

1.5.3.5 Oxidation

Chemical oxidation breaks down organic molecules into biodegradable substances, including ketones, carboxylic acids, alcohols, aldehydes, carbon dioxide, and water. Ozone is a well-known, potent oxidizing agent in chemical oxidation that can be used in chemical interactions with a wide range of inorganic and organic substances [90]. Since no toxic byproducts are produced in the processes involving ozone, the use of this molecule in the treatment of wastewater has become essential during the past decade [91]. Ozonation is a complicated oxidation process that occurs when ozone is added to wastewater and is responsible for a substantial increase in biodegradability. Numerous studies have indicated that ozonation is a highly effective method for removing personal care and pharmaceutical goods since most can be removed. Ozone does have a short half-life, however, and when its levels exceed a particular cutoff point, which is close to 23%, it is viewed as a problem [92]. Recent studies show that ozone has a limited utilization ability and that its oxidative potential to mineralize organic pollutants is ineffective because it also results in the development of hazardous byproducts [93]. If this problem can be somewhat minimized, the catalytic advanced oxidation process has recently been recognized as a workable method for wastewater treatment.

1.5.3.6 Ion exchange

With its high efficiency and low cost, as well as its appealing characteristics for water purification, the treatment of textile effluents, and removal of heavy metals, the ion exchange membrane has garnered significant interest in recent years. The ion exchange technique effectively removes dyes from aqueous solutions owing to the strong interaction between functional groups on ion exchange resins and charged dyes. The categorization of resins may be broken down into four primary groups, which are as follows: cation exchange membrane, anion exchange membrane, cross-linkage membrane, and other ion exchange membrane. The cation and anion membranes are the ones that are utilized the most frequently in the process of dye removal. An anion exchange membrane has positively charged groups, such as $-PR^{3+}$, $-NR_2H^+$, $-NRH^{2+}$, $-NR^{3+}$, and $-NH^{3+}$, which are advantageous because they allow anions to flow through but stop cations from passing through. Negatively charged groups, such as $-C_6H_4O^-$, $-COO^-$, $-PO_3H^-$, $-PO_3^{2-}$, and $-SO_3$ are found in cation exchange membranes [94]. These groups enable cations to flow through the membrane but prevent anions from doing so.

Researchers Muhammad *et al* [95] looked into the effectiveness of removing the anionic dye known as Congo red (CR) from aqueous solutions by employing an anion exchange membrane called an EBTAC. A number of variables, such as contact duration, membrane dose, ionic strength, and temperature, were investigated for their potential impact. According to the findings, the contact duration as well as the membrane do have a substantial influence on the amount of CR that can be removed while the temperature and ionic strength have a negligible influence on the elimination of CR. Akazdam and Chafi continued their research into the removal of Acid Orange 7 dye (AO7) by utilizing a macroporous, highly basic anion exchange resin [96]. The effectiveness of AO7 removal was investigated in a number of trials, with particular attention paid to contact time, pH, solution temperature, agitation rate, adsorbent dose, and starting dye concentration. The authors noted the fact that the pH of the solution has a significant influence on the magnitude of this activity. Within one hundred minutes, the resin could remove as much as 75% of AO7. According to the findings, the experimental data were successfully mapped into the Langmuir isotherm model, and the maximum adsorption capacity was determined to be 200 mg g^{-1}. The adsorption kinetics demonstrated that the adsorption procedure followed a pseudo-second-order model, and thermodynamic analyses further demonstrated that the adsorption of AO7 on the macroporous basic anion exchange resin under the investigated conditions was an exothermic and spontaneous procedure.

1.5.3.7 Precipitation

The chemical precipitation process is an efficient method for removing heavy metals, particularly those found in the discharge produced by the electroplating and paper manufacturing industries. In this process, heavy metals in the contaminated water react with chemical precipitants such as iron, salts, lime, alum, and some polymers to generate insoluble precipitates [97–99]. Metals may be extracted with less effort through the use of this process. Figure 1.5(d) provides a straightforward

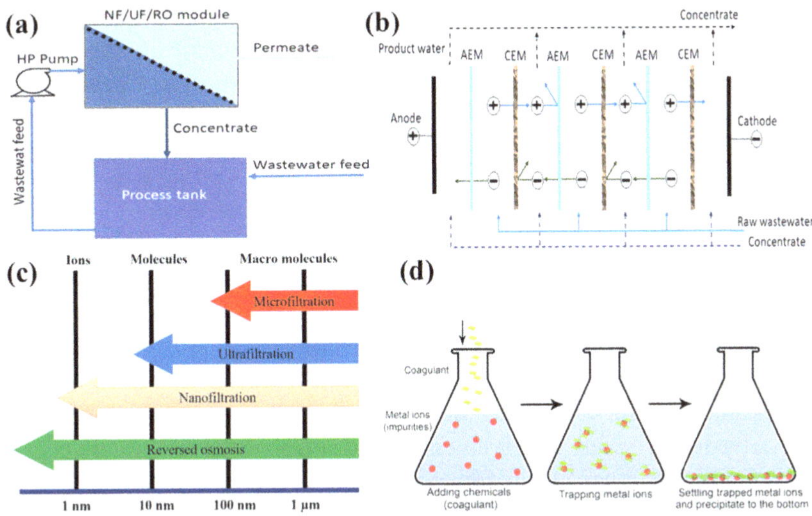

Figure 1.5. (a) Nanofiltration, ultrafiltration, or reverse osmosis method. (b) Electrodialysis method in which alternately charged positive and negative membranes are used. (c) The separation capabilities of different membranes against different pollutants. (d) A simple schematic of the chemical precipitation process. ((a), (b), and (d) Adapted with permission from [76]. Creative Commons Attribution 4.0 International License http://creativecommons.org/licenses/by/4.0/. (c) Reproduced with permission from [112]. Copyright 2022 Elsevier.)

diagrammatic representation of the chemical precipitation process. It is possible to increase the removal capacity and efficiency by adjusting factors such as the temperature, ion charge, starting concentration, and pH [100]. The most significant effect on the treatment is when the pH is adjusted to basic conditions, typically falling between 9 and 11 on the pH scale. The chemical precipitation process may be broken down into two categories: hydroxide precipitation and sulfide precipitation. It is possible that adding coagulants to the hydroxide precipitation process might make it easier to remove heavy metals by sedimentation or filtration. However, in contrast, the sludge generated during the process of metal sulfide precipitation can be removed using gravity settling or filtering. Due to the toxicity of the sulfide ion and hydrogen sulfide (H_2S), this method requires pre-treatment, as well as post-treatment, to accurately regulate the input of reagents.

1.5.3.8 Micro- and ultrafiltration

Microfiltration (MF) removes viruses, bacteria, pollutants, micron-sized particles, contaminants, protozoa, and so on from a solution, solvent, or fluid using a microporous membrane [101]. The MF method is another low-pressure-driven membrane technology; the membrane holes in this process range from 0.1 to 10 μm [101]. MF membranes are constructed out of ceramics, silica, PVC, zirconia, polysulfone, alumina, PTFE, PVDF, polypropylene, polycarbonate, polyamides, cellulose esters, cellulose acetate, or composite materials. The pharmaceutical and biological industries are excellent examples of successful commercial applications of MF. The application of MF systems can also be found in the semiconductor industry for the removal of particles from the sterilization of wine and beer, rinse water, the clarity of other cider and juices, and the treatment of wastewater [101]. Due to the low removal ability of MF, its usage in the removal of heavy metals has not received adequate attention. Furthermore, it has been employed for membrane modification or chemically pre-treating feed solution. Crossflow and dead-end are the two primary configurations that may be used for the MF process, and the mode of application determines their use.

Ultrafiltration (UF) is used as low transmembrane operating pressure (TMP). Due to the possibility that the UF membrane holes could be larger than metal ions, additives can be bound to the metal ions to increase their size. Consequently, it is recommended to use polymer-enhanced ultrafiltration (PEUF) and micellar-enhanced ultrafiltration (MEUF). When a surfactant and UF are combined, MEUF is produced. The strong flux and high selectivity in MEUF result in minimal energy use, reduced space requirements, and high removal effectiveness [102]. Wastewater with low quantities of heavy metals is the best candidate for MEUF treatment [103]. Surfactants are added to wastewater in MEUF at concentrations higher than the critical micellar concentration (CMC). In addition to the CMC, the assembly of surfactant monomers increases the production of certain micelles in the solution. A hydrophilic head and a hydrophobic tail make up the surfactant. The hydrophobic internal core of the micelles can solubilize organic compounds with low molecular weight as a solubilizate, whilst the surface of the micelles can adsorb counter metallic ions on their surface due to electrostatic interactions [104]. Surfactants often attain the maximum retention, the electric charge of which is opposite to that of the metal ions [105–107]. In this context, anionic surfactants, polyelectrolytes (PE), and cationic surfactants are valuable tools for extracting heavy metals [105]. The performance of MEUF depends on a number of elements, such as the solutes supplied, the kind of surfactant employed, the operating circumstances, and the type of membrane utilized.

The combination of UF with binding polymers leads to the formation of PEUF. It is possible for the functional groups of the bonding polymers to be phosphonic, sulfonate, amine, or carboxylated and for them to be connected via ionic or chelating bonds [108]. PEUF is a method that prevents water and other simple components from passing through the pores of a membrane while simultaneously removing polymer-bonded metal ions [109]. PEUF demonstrates excellent and effective extraction, polymer bonding, the capacity to reuse and recover complexation polymer of retentate, low-cost operation, and low energy demands [105, 110]. The choice of appropriate water-soluble polymer macro-ligands is the biggest challenge facing the development of this technology.

Nanofiltration (NF) is a methodology that can concentrate constituents with molecular weights greater than 1000 Da and remove solutes with sizes between 0.0005 and 0.007 μm and molecular weights greater than 200 Da. Therefore, the spectrum of operation for NF is somewhere in between that of UF and the reverse osmosis process. Figures 1.5(a)–(c) provide simplified schematics of several membrane-based filtering procedures.

1.5.3.9 Reverse osmosis

Only molecules less than 0.5–1.5 nm in size can penetrate through the semi-permeable membrane used in the reverse osmosis (RO) separation process, which is a pressure-driven separation method. The RO process works by delivering a pressure between 20 and 70 bar higher than the osmotic pressure of the feed solution. This causes the usual osmosis process to be inverted [101]. Typically, the particle mass of the solutes that are inhibited falls anywhere between 0.000 25 and 0.003 μm. The RO process had the potential to remove between 95% and 99% of the charged organic and inorganic salts. The RO process is both space-efficient and capable of achieving high rejection efficiencies. However, the greatest drawback of RO systems is their propensity for membrane fouling and breakdown [101]. With a removal efficiency of greater than 98.7562%, the RO separation process was utilized to recover heavy metal ions from electroplating wastewater. These ions included Ni^{2+}, Cr^{6+}, and Cu^{2+}. RO has recently been utilized to filter industrial wastewater from costar-field mining activities in Australia (Victoria) [111]. The mean extraction efficiency for Zn^{2+}, Fe^{3+}, Ni^{2+}, Sb^{3+}, and $As^{3+}$was, respectively, 48%, 10%, 82%, 95%, and 66% [111].

1.5.3.10 Electrodialysis

The procedure known as electrodialysis (ED) employs an electrical potential difference to separate ions. An optional parallel arrangement of anion exchange membranes (AEM) and cation exchange membranes (CEM) is used in ED to separate ionic solutes [110]. The cations are carried by the CEM, whereas the anions are transferred through the AEM during the ED process. As illustrated in figure 1.5(c), in this scenario half of the channels in the ED stack create the treated stream (dilute), and the other half discharge the concentrated stream. ED has a high water-recovery rate, does not entail any phase shift, reaction, or chemical interaction [113], and is capable of functioning across a broad pH spectrum. On the other hand, ED is characterized by the high cost of membranes, membrane fouling, and a necessity for an electrical potential.

Using a novel ED heterogeneous CEM (consisting of 2-acrylamide-2-methyl propane sulfonic acid-based hydrogel and PVC), ED has been used to separate K^+, Ni^{2+}, and Pb^{2+} from synthetic solution [114]. This has resulted in an extraction efficiency of 99.9%, 96.9%, and 99.9% for K^+, Ni^{2+}, and Pb^{2+}, respectively. In order to recover Pb^{2+}, a batch ED technique was utilized, and the highest possible separation efficiency was achieved (100%) [115]. Additionally, a pilot-scale ED system has been utilized to remove Ni^{2+}, Cu^{2+}, and traces of Fe^{3+}, Cd^{2+}, Zn^{2+}, and Cr^{6+}, and the removal rates have approached 90% [116]. ED was used to extract As^{5+} and As^{3+} from metallurgical effluent, and the overall removal effectiveness was determined to be 91.38% [117].

1.5.3.11 Adsorption

The process by which liquid or gas droplets can concentrate on a porous object's surface is known as adsorption. Adsorbate is the term used to describe the material that accumulates at the interface, while adsorbent refers to the substance that comes into contact with the solid. When considering the manner in which the adsorbate species is deposited onto the adsorbent surface, the adsorption process can often be split into two distinct categories: chemical sorption and physical sorption. During the process of physical adsorption, the adsorbate molecules bind to the adsorbents by applying weak physical forces, such as those governed by hydrophobicity, van der Waals, polarity, static interactions, hydrogen bonding, π–π interactions, and dipole interactions. Through the process of chemical adsorption, adsorbates are electrically and chemically attached to the adsorbent surface through the electron exchange process. An adsorption is a popular approach for removing contaminants that offers an appealing alternative. This is because, compared to other approaches, it has been found to be superior in terms of simplicity of design, flexibility, cost, and high efficiency for removing dyes from wastewater. In addition, adsorption does not result in the production of dangerous compounds and prevents the occurrence of secondary contamination. The surface area, adsorption capacity, and potential for reuse cost of the materials were considered in the selection of an environmentally acceptable as well as inexpensive adsorbent to employ in the adsorption process [118].

The most popular heavy metal adsorbent is activated carbon (AC), which can be produced by carbonizing or pyrolyzing various materials containing carbon and then activating the resulting carbon. When employing waste from the agricultural business as a precursor for AC, adsorption capacities of up to roughly 900 mg g^{-1} have been achieved for the elimination of Cr^{3+} from water [119]. Various raw materials, processing variables, and removal conditions have been researched, including agricultural waste and petroleum pitches. This research aims to find ways to lower the treatment cost of using AC by increasing its recyclability, increasing its operational life, developing environmentally friendly activation processes, and improving its adsorption capacity, to name just a few of these improvements [120–122]. Figure 1.6 provides a graphical depiction of the adsorptive removal of dye pollution in water by employing various adsorbents.

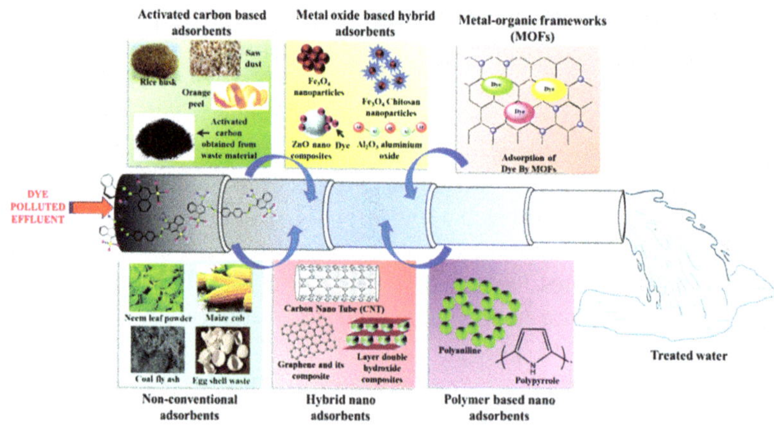

Figure 1.6. A schematic representation of the water pollution caused by dye wastewater, its removal via adsorption utilizing a variety of adsorbents, and the production of clean water. (Reproduced from [123]. Attribution 3.0 Unported (CC BY 3.0) https://creativecommons.org/licenses/by/3.0/.)

1.5.3.12 Electrolysis

Electrolysis is a process that can either deposit or decompose soluble materials on an electrode surface through the use of electrochemical redox reactions. In this process, most of the metal ions are deposited on the electrode surface, while the organic compounds are broken down into water and carbon dioxide, as well as other products that are either low in toxicity or non-toxic [124, 125]. In addition to that, it has also been utilized in the process of removing color and turbidity from polluted water. This approach is efficient for removing total dissolved solids (TDSs) levels lower than 200 mg l^{-1}, and as a result, the wastewater must be pre-treated before it can be processed further. The temperature, amount of current, applied pH, and contact duration are the variables that determine the outcome of the process. Electrolysis, as a method for treating wastewater, has not yet reached its full potential and is currently being developed as a technology. It has very infrequently been utilized on a commercial scale for the treatment of wastewater. However, one of its uses is the treatment of some industrial effluents, particularly those abundant in metal ions and certain organic compounds. It is possible to employ this method as a water source reduction strategy. The deposited metal ions may be used in additional applications thanks to this method, which eliminates the need for any waste management concerns to be addressed.

1.5.3.13 Advanced oxidation process

The advanced oxidation process (AOP) has emerged in recent years as a potentially promising approach for the treatment of polluted water that contains rigid organic molecules [126]. This is a result of the capability of this mechanism to make use of reactive radicals produced by hydroxyl species. The source of the reactive radicals differs depending on the sort of advanced oxidation technique being utilized. When degrading organic chemicals in wastewater, the conventional or straightforward oxidation process does not consistently achieve its intended function. As a result, a strategy known as advanced oxidation has been developed, in which several simultaneous oxidation processes are carried out with the assistance of highly reactive species of hydroxyl molecules acting in the capacity of free radicals. Some recent oxidation methods utilized in wastewater treatment include photocatalysis, solar-driven processes, and photo-Fenton [127].

The utilization of photocatalysis as a treatment strategy for polluted water shows great promise. Because this technique takes advantage of sunshine, it is not only economical but also beneficial to the environment and universally adaptable. Photocatalysis may decompose many organic substances, including dyes, crude oils, microorganisms, and inorganic compounds such as nitrous oxide. Additionally, this process can be used with precipitation and filtering to remove heavy metals [128]. This technology using nano-based catalysts is utilized not only for cleaning water but also for preventing air pollution and creating construction materials with self-cleaning surfaces. Titanium oxide, often known as TiO_2, is the photocatalyst that sees the most widespread application. It is activated by ultraviolet (UV) irradiation, which can come from natural or man-made sources. The elimination of phenolic chemicals is the primary application for this technology, which is employed in the oil industry. TiO_2 is a photocatalytic semiconductor material that

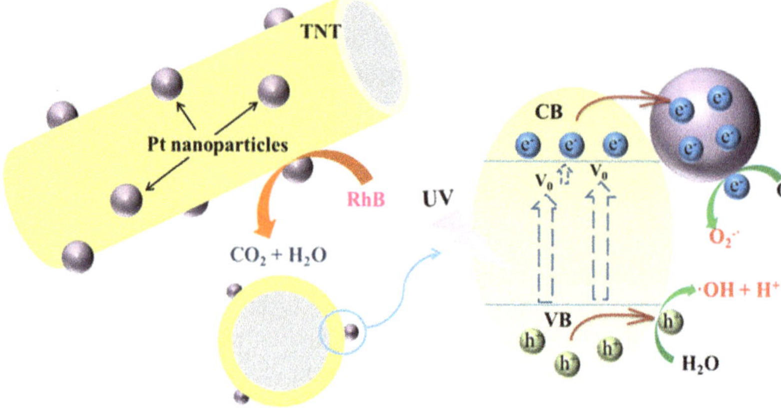

Figure 1.7. A potential mechanism for the deterioration of methyl benzoate imparted by Pt–TiO_2 nanotubes. (Reproduced from [130]. Creative Commons Attribution 4.0 International License http://creativecommons.org/licenses/by/4.0/.)

reduces and eliminates these hazardous and toxic chemicals from polluted water. The treatment of generated water and wastewater outflow from garage forecourts may be accomplished with this extremely practicable approach [129]. However, one of the most significant drawbacks of this technique is that it is not very effective at the adsorption and treatment of highly concentrated organic pollutants using electricity. Figure 1.7 shows a possible schematic mechanism of photocatalysis.

The photo-Fenton process is the result of a combination of the Fenton reaction and exposure to UV light [131]. As a result of the creation of a hydroxyl radical, the synthesis of secondary chlorinated products is restricted in this process, which is one of the reasons why this technique is considered to be one of the most effective methods for water purification. Studies have shown that this method of wastewater treatment is not only straightforward in terms of technology but also affordable. Additionally, it has a high pollutant removal effectiveness, particularly with regard to antibiotics, making it an attractive option. Compared to the traditional Fenton process, the photo-Fenton method has a lower consumption of hydrogen peroxide due to the increased rate of hydroxyl radical formation caused by the application of UV light [132].

1.6 Influencing factors for wastewater treatment

We must examine the factors that affect the removal of each metal from water because each one has unique behaviors and properties. The following sections summarize the effects of temperature, pH, natural organic matter concentration, and ionic strength on removing heavy metals from household wastewater effluents, lakes, rivers, and groundwater.

1.6.1 pH

pH changes can impact the elimination of the current contaminants as they may change the balance of inorganic and organic compounds in water. These alterations may result from added chemicals during the restoration process or may be rooted in biological instability [133]. When the pH value is neutral or below, heavy metals enter a state called the cationic state, which causes them to be mobile and more soluble in water. Higher pH levels, on the other hand, affect their oxidation state and encourage the development of hydroxides as well as other anions [134]. For instance, the pH of the solution affects both the removal efficiency of natural clay and its adsorption capacity when employed to eliminate Ni^{2+}. The effectiveness of nickel removal rose in tandem with the pH of the solution [135].

1.6.2 Temperature

There is better efficacy in elimination when the temperature is higher, according to tests carried out using a variety of treatments [136]. When the temperature, for instance, increased to its highest point of 55 °C, the effectiveness of Cd^{2+}, Cu^{2+}, Pb^{2+}, and Ni^{2+} removal from the environment employing plants rose [137]. At four different temperatures (22 °C, 30 °C, 35 °C, and 40 °C), researchers looked at how temperature affected the ability of hydrogels to adsorb copper (Cu^{2+}) and lead (Pb^{2+}) ions, respectively. The optimal temperature for Pb^{2+} was 30 °C, while the optimal temperature for Cu^{2+} was 35 °C. An increased temperature of up to 40 °C produced a better removal result [138]. When the temperature was increased from 5 °C to 40 °C, there was a discernible improvement in the effectiveness of pistachio residues in eliminating Cr^{6+}. This behavior was attributed to increased adsorption sites on the surface of the material [139]. Additionally, whenever the temperature increased from 10 °C to 60 °C in a study employing fungi biomass, the biosorption effectiveness of heavy metals rose. This was explained by a greater affinity of the active sites of the biomass substrates [140].

On the other hand, the eradication of heavy metals might be negatively impacted by even slight temperature variations. For instance, when magnetic iron oxide nanoparticles were used as adsorbents, the removal efficiency of Ni^{2+}, Cu^{2+}, and Al^{3+} decreased as the temperature increased. This is because the chemisorption rate at higher temperatures prevented heavy metals from reaching new active sites on the particle surface, which would have boosted their capacity to remove metals. Additionally, the temperature has the ability to alter the nanoparticle pores size, which results in a diminished adsorption capacity [141].

1.6.3 Ionic strength

The overall concentration of ions presents inside the aqueous solution, along with their chemical charges, determine the ionic strength. This strength affects the removal of heavy metals and the ion exchange process between metal ions and adsorbents. For instance, when sodium chloride (NaCl) concentration was raised, the adsorption of ions such as Cr^{3+}, Pb^{2+}, Zn^{2+}, and Cu^{2+} on porous magnetic particles decreased somewhat [142]. In another study, the amount of Cu^{2+} and Ni^{2+} that sticks to bentonite goes up as the ionic strength decreases at pH 5.5, with the maximum amount of sorption occurring when the NaCl concentration is less than 0.05 M [143]. This is due to the fact that greater ionic strength might lessen the molecules' electrostatic repulsion, causing the particles to aggregate more and diminish the system's accessible binding sites and sorption. Furthermore, the production of soluble metal chloride complexes, which are challenging to remove from water, might be aided by the existence of chloride [144]. This is due to the low adsorption affinity of the formed chloride complexes [145].

1.6.4 Natural organic matter

As a result of interactions between the biosphere and the geosphere (e.g. decomposition of plants and animals in water), natural organic matter (NOM) can be found in various water sources, although it is most prevalent in surface waters. As a consequence, NOM can engage in chemical reactions with heavy metals found in polluted water through processes including adsorption, chelation, and ion exchange. The presence of NOM affects the reactivity, mobility, bioavailability, and toxicity of heavy metals, despite the fact that it is hard to precisely quantify how much of an influence it has on eliminating heavy metals. For example, lead may combine with naturally occurring microorganisms in natural waterways or wastewater to create complexes, making the lead more mobile and challenging to remove using traditional techniques, including coagulation/flocculation and chemical precipitation. According to a study using titanate nanotubes supported by powder-activated charcoal, complex formation and weak interactions between the nanotubes and dissolved organic matter affect the capacity of Pb^{2+} adsorption [146]. Because of complex formation, simultaneous chelation, and ionic competition between metal and NOM, high concentrations of NOM can make it more difficult for heavy metals to be absorbed and removed from the body [147]. These characteristics and their interactions present substantial obstacles for heavy metals removal from polluted water. Furthermore, because they may impact repeatability, it is important to consider them while evaluating different removal methods.

1.7 Selection of wastewater technologies

The characteristics of the wastewater being treated and the conditions that must be met (including economic aspects) will guide the choice of treatment technology. After completing the primary and secondary water treatment procedures, highly contaminated water that is cloudy and contains solid debris is subjected to advanced tertiary (chemical) water treatment technologies. If the BOD is insignificant, there is no need for a further step. Only tertiary water treatment is necessary if the water is completely clear, does not include any solids, and is contaminated by biological, organic, and inorganic contaminants. In most cases, groundwater is contaminated by toxic metal anions and ions, and the only technologies that can effectively treat this pollution are tertiary water technologies. On the other hand, treating surface water polluted with biological, organic, and inorganic contaminants requires secondary and tertiary levels of treatment. In most cases, wastewater is extremely polluted, and it is possible for it to be colored with solid waste that contains biological, organic, and inorganic pollutants. Because of this, it is necessary to employ a treatment method that successfully combines physical, biological, and

Table 1.1. Disadvantages and advantages of wastewater treatment techniques along with their cost.

Methods	Advantages	Disadvantages	Cost (US$ per million liters)
Screening	Minimizes treatment technology interruption and obstruction.	Ineffective for emerging contaminants removal.	25–450
Sedimentation	Avoids clogging, removes dirt and solids.	Effective for municipal or domestic sewage.	5–10
Coagulation	Cost-effective and easy to operate.	Ineffective for the removal of heavy metals. Sludge production. Utilization of expensive chemicals.	25–500
Floatation	More efficient than screening, sedimentation, and skimming. Eliminates several emerging contaminants quickly. Encourages hybrid wastewater treatment.	Ineffective at diminishing emerging contaminants. Very costly and complex.	5–25
Aerobic processes	Removes several emerging contaminants. Can be combined well with other treatment methods.	Long hydraulic period and sludge retention. Requires vast amounts of land. Expensive.	20–200
Anaerobic process	Converts wastes into biogas without oxygen.	Insufficient to eliminate inorganic contaminants.	20–200
Adsorption	Removes emerging contaminants precisely. Fast and reliable elimination can aid subsequent treatment methods.	Cyanotoxin accumulation in the adsorbent. Adsorbents are highly specific, which makes it difficult to remove unknown toxins.	50–150

(*Continued*)

Table 1.1. *(Continued)*

Methods	Advantages	Disadvantages	Cost (US$ per million liters)
Precipitation	Highly sufficient metal and fluoride removal. Simplicity. Physicochemical integrated process. Economically viable and effective. Can remove heavy pollutants. Low chemical oxygen demand.	Consumption of chemicals. Unproductive in metal ion elimination at small concentrations. Production of large amounts of sludge. For the elimination of complex metals, an oxidation step is needed.	20–500
Ion exchange	Eliminates pollutants quickly. Treats inorganic waste effectively. Quick. Uses inexpensive materials.	pH sensitive. Metal ion fouling. Suitable only for metals with low concentrations. High operational cost.	50–200
Electrochemical technologies	Eco-friendly and simple. Labor- and energy-saving.	Huge capital expenditure. Requires costly electricity. Unhelpful at low aqueous metal ion concentrations.	For electrodialysis 15–400
Membrane technologies	Contaminant removal efficiency is maximum. Less energy consumption. Free from chemical additives, phase change, and secondary pollution. Easy to fabricate. Can remove both inorganic and organic compounds.	The process is complex. Very costly. Fouling of membrane. Permeate flux is lower.	For micro- and ultrafiltration 15–200, reverse osmosis 20–450

Advanced oxidation process	Reactive radicals are locally produced. Lower chemical consumption. Pollutant mineralization. Free from sludge production. The degradation process is quick. Effective for dyes, drugs, etc.	The process works under a laboratory scale. Unprofitable for businesses of all sizes. Byproduct formation.	—
Evaporation	Energy-saving. Efficient processes	High energy consumption for high volumes of polluted water. Expensive for small-scale industries. Causes a high pollution load.	15–200
Oxidation	Extremely effective at eliminating personal care and pharmaceutical products.	Utilization efficiency is low. The ability of mineralizing chemicals to oxidize. Byproduct formation.	100–2000

chemical treatment methodologies. The sorts of contaminants found in the water should be considered when deciding which tertiary water treatment technologies to use, and the best choice can be made by considering the topics presented earlier. The advantages and disadvantages of the aforementioned water remediation techniques with treatment cost (US$ per million liters) are summarized in table 1.1.

References

[1] WHO/UNICEF 2015 *Progress on Sanitation and Drinking Water: 2015 Update and MDG Assessment* https://apps.who.int/iris/handle/10665/177752 (Accessed: 12 September 2022)

[2] Schwarzenbach R P, Escher B I, Fenner K, Hofstetter T B, Johnson C A, Von Gunten U and Wehrli B 2006 The challenge of micropollutants in aquatic systems *Science* **313** 1072–7

[3] Vörösmarty C J *et al* 2010 Global threats to human water security and river biodiversity *Nature* **467** 555–61

[4] Ewaid S H and Abed S A 2017 Water quality index for Al-Gharraf River, southern Iraq *Egypt. J. Aquat. Res.* **43** 117–22

[5] Ewaid S H, Abed S A and Al-Ansari N 2019 Crop water requirements and irrigation schedules for some major crops in southern Iraq *Water* **11** 756

[6] Chowdhary P, Bharagava R N, Mishra S and Khan N 2020 Role of industries in water scarcity and its adverse effects on environment and human health *Environmental Concerns and Sustainable Development* (Singapore: Springer) pp 235–56

[7] Cheng H, Chen C, Wu S, Mirza Z A and Liu Z 2017 Emergy evaluation of cropping, poultry rearing, and fish raising systems in the drawdown zone of Three Gorges Reservoir of China *J. Clean. Prod.* **144** 559–71

[8] Wu H, Gai Z, Guo Y, Li Y, Hao Y and Lu Z N 2020 Does environmental pollution inhibit urbanization in China? A new perspective through residents' medical and health costs *Environ. Res.* **182** 109128

[9] De Zuane J 1990 *Handbook of Drinking Water Quality: Standards and Controls* (New York: Van Nostrand Reinhold)

[10] Rathi B S, Kumar P S and Vo D V N 2021 Critical review on hazardous pollutants in water environment: occurrence, monitoring, fate, removal technologies and risk assessment *Sci. Total Environ.* **797** 149134

[11] Parris K 2011 Impact of agriculture on water pollution in OECD countries: recent trends and future prospects *Int. J. Water Resour. Dev.* **27** 33–52

[12] Moss B 2008 Water pollution by agriculture *Philos. Trans. R. Soc.* B **363** 659–66

[13] Lu Y *et al* 2015 Impacts of soil and water pollution on food safety and health risks in China *Environ. Int.* **77** 5–15

[14] Lai W 2017 Pesticide use and health outcomes: evidence from agricultural water pollution in China *J. Environ. Econ. Manage.* **86** 93–120

[15] Xiao J, Wang L, Deng L and Jin Z 2019 Characteristics, sources, water quality and health risk assessment of trace elements in river water and well water in the Chinese Loess Plateau *Sci. Total Environ.* **650** 2004–12

[16] Hanif M, Miah R, Islam M and Marzia S 2020 Impact of Kapotaksha river water pollution on human health and environment *Prog. Agric.* **31** 1–9

[17] Brown J and Clasen T 2012 High adherence is necessary to realize health gains from water quality interventions *PLoS One* **7** e36735

[18] Halder J and Islam N 2015 Water pollution and its impact on the human health *J. Environ. Hum.* **2** 36–46
[19] Dwivedi S, Mishra S and Tripathi R D 2018 Ganga water pollution: a potential health threat to inhabitants of Ganga basin *Environ. Int.* **117** 327–38
[20] Ustaoğlu F, Tepe Y and Taş B 2020 Assessment of stream quality and health risk in a subtropical Turkey river system: a combined approach using statistical analysis and water quality index *Ecol. Indic.* **113** 105815
[21] Kamble S M 2014 Water pollution and public health issues in Kolhapur City in Maharashtra *Int. J. Sci. Res. Publ.* **4** 1–6
[22] Halder J and Islam N 2015 Water pollution and its impact on the human health *J. Environ. Hum.* **2** 36–46
[23] Bhangaonkar P D and Patel J S 2019 Assessment of heavy metals in surface water of Vishwamitri River *Int. J. Hydrol. Sci. Technol.* **9** 675–89
[24] Krishnan S and Indu R 2017 Groundwater contamination in India: discussing physical processes, health and socio-behavioural dimensions *BMC Public Health* **5** 1–8
[25] Jabeen S, Mahmood Q, Tariq S, Nawab B and Elahi N 2011 Health impact caused by poor water and sanitation in district Abbottabad *J. Ayub Med. Coll. Abbottabad* **23** 47–50
[26] Currie J, Graff Zivin J, Meckel K, Neidell M and Schlenker W 2013 Something in the water: contaminated drinking water and infant health *Can. J. Econ.* **46** 791–810
[27] Sharma B 2016 Environmental pollution: its effects on life and its remedies *Biochem. Cell. Arch.* **16** 57–60
[28] Ahmed T, Scholz M, Al-Faraj F and Niaz W 2016 Water-related impacts of climate change on agriculture and subsequently on public health: a review for generalists with particular reference to Pakistan *Int. J. Environ. Res. Public Health* **13** 1051
[29] Chowdhury S, Krause-Pilatus A and Zimmermann K F 2016 Arsenic contamination of drinking water and mental health *Discussion Papers on Development Policy* No. 222 SSRN
[30] Bolong N, Ismail A F, Salim M R and Matsuura T 2009 A review of the effects of emerging contaminants in wastewater and options for their removal *Desalination* **239** 229–46
[31] Templeton M R, Graham N and Voulvoulis N 2009 Emerging chemical contaminants in water and wastewater *Philos. Trans. R. Soc.* A **367** 3873–5
[32] Dewitt J C, Peden-Adams M M, Keller J M and Germolec D R 2012 Immunotoxicity of perfluorinated compounds: recent developments *Toxicol. Pathol.* **40** 300–11
[33] Lee I and Viberg H 2013 A single neonatal exposure to perfluorohexane sulfonate (PFHxS) affects the levels of important neuroproteins in the developing mouse brain *Neurotoxicology* **37** 190–6
[34] Bonefeld-Jorgensen E C *et al* 2011 Perfluorinated compounds are related to breast cancer risk in Greenlandic Inuit: a case control study *Environ. Health* **10** 88
[35] Okoh A I, Sibanda T and Gusha S S 2010 Inadequately treated wastewater as a source of human enteric viruses in the environment *Int. J. Environ. Res. Public Health* **7** 2620–37
[36] Craun G F, Brunkard J M, Yoder J S, Roberts V A, Carpenter J, Wade T, Calderon R L, Roberts J M, Beach M J and Roy S L 2010 Causes of outbreaks associated with drinking water in the United States from 1971 to 2006 *Clin. Microbiol. Rev.* **23** 507–28
[37] Waddington H, Snilstveit B, White H and Fewtrell L 2009 Water, sanitation and hygiene interbentions to combat childhood diarrhoea in developing countries *Systematic Review* 001 International Initiative for Impact Evaluation https://www.3ieimpact.org/sites/default/files/2019-01/sr1-WASH.pdf

[38] Ahmed S and Ismail S 2018 Water pollution and its sources, effects and management: a case study of Delhi *Int. J. Curr. Adv. Res.* **7** 10436–42

[39] Ansari Z Z and Akhmatov S 2020 *Impacts of Water Pollution on Human Health: A Case Study of Delhi* (in Russian) 163–8

[40] Clasen T F, Alexander K T, Sinclair D, Boisson S, Peletz R, Chang H H, Majorin F and Cairncross S 2015 Interventions to improve water quality for preventing diarrhoea *Cochrane Database Syst. Rev.* **2015** CD004794

[41] Khan S, Shahnaz M, Jehan N, Rehman S, Shah M T and Din I 2013 Drinking water quality and human health risk in Charsadda district, Pakistan *J. Clean. Prod.* **60** 93–101

[42] Fewtrell L and Colford J M 2005 Water, sanitation and hygiene in developing countries: interventions and diarrhoea-a review *Water Sci. Technol.* **52** 133–42

[43] Arnold B F and Colford J M 2007 Treating water with chlorine at point-of-use to improve water quality and reduce child diarrhea in developing countries: a systematic review and meta-analysis *Am. J. Trop. Med. Hyg.* **76** 354–64

[44] Kazi T G, Arain M B, Baig J A, Jamali M K, Afridi H I, Jalbani N, Sarfraz R A, Shah A Q and Niaz A 2009 The correlation of arsenic levels in drinking water with the biological samples of skin disorders *Sci. Total Environ.* **407** 1019–26

[45] Hanif M, Miah R, Islam M and Marzia S 2020 Impact of Kapotaksha river water pollution on human health and environment *Prog. Agric.* **31** 1–9

[46] Arif A 2020 Water pollution and industries *Pure Appl. Biol.* **9** 2214–24

[47] Yau V, Wade T J, de Wilde C K and Colford J M 2009 Skin-related symptoms following exposure to recreational water: a systematic review and meta-analysis *Water Qual. Expo. Heal.* **1** 79–103

[48] Fleisher J M and Kay D 2006 Risk perception bias, self-reporting of illness, and the validity of reported results in an epidemiologic study of recreational water associated illnesses *Mar. Pollut. Bull.* **52** 264–8

[49] Glade M J 1999 Food, nutrition, and the prevention of cancer: a global perspective *Nutrition* **15** 523–6

[50] Lin H J, Sung T I, Chen C Y and Guo H R 2013 Arsenic levels in drinking water and mortality of liver cancer in Taiwan *J. Hazard. Mater.* **262** 1132–8

[51] Ferreccio C, González C, Milosavjlevic V, Marshall G, Sancha A M and Smith A H 2000 Lung cancer and arsenic concentrations in drinking water in Chile *Epidemiology* **11** 673–9

[52] Schullehner J, Hansen B, Thygesen M, Pedersen C B and Sigsgaard T 2018 Nitrate in drinking water and colorectal cancer risk: a nationwide population-based cohort study *Int. J. Cancer* **143** 73–9

[53] Zhitkovich A 2011 Chromium in drinking water: sources, metabolism, and cancer risks *Chem. Res. Toxicol.* **24** 1617–29

[54] Tseng C H, Lei C and Chen Y C 2018 Evaluating the health costs of oral hexavalent chromium exposure from water pollution: a case study in Taiwan *J. Clean. Prod.* **172** 819–26

[55] Marmot M *et al* 2007 *Food, Nutrition, Physical Activity, and the Prevention of Cancer: a Global Perspective* (Washington, DC: American Institute of Cancer Research)

[56] Zhang X L, Zhang B, Zhang X, Chen Z F, Zhang J Z, Liang S Y, Men F S, Zheng S L, Li X P and Bai X L 2003 Research and control of well water pollution in high esophageal cancer areas *World J. Gastroenterol.* **9** 1187–90

[57] Xu C, Xing D, Wang J and Xiao G 2019 The lag effect of water pollution on the mortality rate for esophageal cancer in a rapidly industrialized region in China *Environ. Sci. Pollut. Res.* **26** 32852–8

[58] Ebenstein A 2012 The consequences of industrialization: evidence from water pollution and digestive cancers in China *Rev. Econ. Stat.* **94** 186–201

[59] Meerbergen K, Willems K A, Dewil R, Van Impe J, Appels L and Lievens B 2018 Isolation and screening of bacterial isolates from wastewater treatment plants to decolorize azo dyes *J. Biosci. Bioeng.* **125** 448–56

[60] Hoiberg B and Shah M T 2021 CFD study of multiphase flow in aerated grit tank *J. Water Process Eng.* **39** 101698

[61] Bennett G 1994 Industrial and hazardous waste treatment *J. Hazard. Mater.* **37** 378

[62] Metcalf L, Eddy H and Tchobanoglous G 1991 *Wastewater Engineering: Treatment, Disposal, and Reuse* (New York: McGraw-Hill)

[63] Chu B T T, Petrovich M L, Chaudhary A, Wright D, Murphy B, Wells G and Poretsky R 2018 Metagenomics reveals the impact of wastewater treatment plants on the dispersal of microorganisms and genes in aquatic sediments *Appl. Environ. Microbiol.* **84** e02168-17

[64] Teh C Y, Budiman P M, Shak K P Y and Wu T Y 2016 Recent advancement of coagulation–flocculation and its application in wastewater treatment *Ind. Eng. Chem. Res.* **55** 4363–89

[65] Hargreaves A J, Vale P, Whelan J, Alibardi L, Constantino C, Dotro G, Cartmell E and Campo P 2018 Correction to: Coagulation–flocculation process with metal salts, synthetic polymers and biopolymers for the removal of trace metals (Cu, Pb, Ni, Zn) from municipal wastewater *Clean Technol. Environ. Policy* **20** 1119

[66] Pang F M, Teng S P, Teng T T and Omar A K M 2009 Heavy metals removal by hydroxide precipitation and coagulation–flocculation methods from aqueous solutions *Water Qual. Res. J. Canada* **44** 174–82

[67] Xu D, Zhou B and Yuan R 2019 Optimization of coagulation–flocculation treatment of wastewater containing Zn(II) and Cr(VI) *IOP Conf. Ser.: Earth Environ. Sci.* **227** 052049

[68] Ibarra-Rodríguez D, Lizardi-Mendoza J, López-Maldonado E A and Oropeza-Guzmán M T 2017 Capacity of 'nopal' pectin as a dual coagulant–flocculant agent for heavy metals removal *Chem. Eng. J.* **323** 19–28

[69] Nourani M, Baghdadi M, Javan M and Bidhendi G N 2016 Production of a biodegradable flocculant from cotton and evaluation of its performance in coagulation–flocculation of kaolin clay suspension: optimization through response surface methodology (RSM) *J. Environ. Chem. Eng.* **4** 1996–2003

[70] Teh C Y, Budiman P M, Shak K P Y and Wu T Y 2016 Recent advancement of coagulation–flocculation and its application in wastewater treatment *Ind. Eng. Chem. Res.* **55** 4363–89

[71] Edzwald J K 2010 Dissolved air flotation and me *Water Res.* **44** 2077–106

[72] Peng W, Chang L, Li P, Han G, Huang Y and Cao Y 2019 An overview on the surfactants used in ion flotation *J. Mol. Liq.* **286** 110955

[73] Hoseinian F S, Rezai B, Kowsari E, Chinnappan A and Ramakrishna S 2020 Synthesis and characterization of a novel nanocollector for the removal of nickel ions from synthetic wastewater using ion flotation *Sep. Purif. Technol.* **240** 116639

[74] Salmani M H, Davoodi M, Ehrampoush M H, Ghaneian M T and Fallahzadah M H 2013 Removal of cadmium (II) from simulated wastewater by ion flotation technique *Iran. J. Environ. Heal. Sci. Eng.* **10** 16

[75] Mahne E J and Pinfold T A 2007 Precipitate flotation I. Removal of nickel from dilute aqueous solutions and its separation from cobalt *J. Appl. Chem.* **18** 52–4

[76] Qasem N A A, Mohammed R H and Lawal D U 2021 Removal of heavy metal ions from wastewater: a comprehensive and critical review *npj Clean Water* **4** 36

[77] Raju S, Carbery M, Kuttykattil A, Senthirajah K, Lundmark A, Rogers Z, SCB S, Evans G and Palanisami T 2020 Improved methodology to determine the fate and transport of microplastics in a secondary wastewater treatment plant *Water Res.* **173** 115549

[78] Kumar R, Sarmah A K and Padhye L P 2019 Fate of pharmaceuticals and personal care products in a wastewater treatment plant with parallel secondary wastewater treatment train *J. Environ. Manage.* **233** 649–59

[79] Barragán B E, Costa C and Carmen Márquez M 2007 Biodegradation of azo dyes by bacteria inoculated on solid media *Dyes Pigm.* **75** 73–81

[80] Zhang H *et al* 2019 Geographical patterns of nirS gene abundance and nirS-type denitrifying bacterial community associated with activated sludge from different wastewater treatment plants *Microb. Ecol.* **77** 304–16

[81] Sepehri A, Sarrafzadeh M H and Avateffazeli M 2020 Interaction between *Chlorella vulgaris* and nitrifying-enriched activated sludge in the treatment of wastewater with low C/N ratio *J. Clean. Prod.* **247** 119164

[82] Thatai S, Verma R, Khurana P, Goel P and Kumar D 2018 Water quality standards, its pollution and treatment methods *A New Generation Material Graphene: Applications in Water Technology* (Cham: Springer) pp 21–42

[83] Fux C, Boehler M, Huber P, Brunner I and Siegrist H 2002 Biological treatment of ammonium-rich wastewater by partial nitritation and subsequent anaerobic ammonium oxidation (anammox) in a pilot plant *J. Biotechnol.* **99** 295–306

[84] Van Der Zee F P and Villaverde S 2005 Combined anaerobic–aerobic treatment of azo dyes —a short review of bioreactor studies *Water Res.* **39** 1425–40

[85] Bernard O, Hadj-Sadok Z, Dochain D, Genovesi A and Steyer J P 2001 Dynamical model development and parameter identification for an anaerobic wastewater treatment process *Biotechnol. Bioeng.* **75** 424–38

[86] Bernet N, Delgenes N, Akunna J C, Delgenes J P and Moletta R 2000 Combined anaerobic–aerobic SBR for the treatment of piggery wastewater *Water Res.* **34** 611–9

[87] Talarposhti A M, Donnelly T and Anderson G K 2001 Colour removal from a simulated dye wastewater using a two-phase anaerobic packed bed reactor *Water Res.* **35** 425–32

[88] Venkata Mohan S, Lalit Babu V and Sarma P N 2007 Anaerobic biohydrogen production from dairy wastewater treatment in sequencing batch reactor (AnSBR): effect of organic loading rate *Enzyme Microb. Technol.* **41** 506–15

[89] Gupta V K, Ali I, Saleh T A, Nayak A and Agarwal S 2012 Chemical treatment technologies for waste-water recycling—an overview *RSC Adv.* **2** 6380–8

[90] Coca M, González-Benito G and García-Cubero M T 2016 Chemical oxidation with ozone as an efficient pretreatment of lignocellulosic materials *Biomass Fractionation Technologies for a Lignocellulosic Feedstock Based Biorefinery* (Amsterdam: Elsevier) pp 409–29

[91] Cuerda-Correa E M, Alexandre-Franco M F and Fernández-González C 2020 Advanced oxidation processes for the removal of antibiotics from water. An overview *Water* **12** 102

[92] Shen J, Ding T and Zhang M 2019 Analytical techniques and challenges for removal of pharmaceuticals and personal care products in water *Pharmaceuticals and Personal Care Products: Waste Management and Treatment Technology: Emerging Contaminants and Micro Pollutants* (Amsterdam: Elsevier) pp 239–57

[93] Wang J and Chen H 2020 Catalytic ozonation for water and wastewater treatment: recent advances and perspective *Sci. Total Environ.* **704** 135249
[94] Xu T 2005 Ion exchange membranes: state of their development and perspective *J. Memb. Sci.* **263** 1–29
[95] Khan M I, Akhtar S, Zafar S, Shaheen A, Khan M A, Luque R and Ur Rehman A 2015 Removal of Congo red from aqueous solution by anion exchange membrane (EBTAC): adsorption kinetics and themodynamics *Materials* **8** 4147–61
[96] Akazdam S, Chafi M, Yassine W and Gourich B 2017 Removal of Acid Orange 7 dye from aqueous solution using the exchange resin Amberlite FPA-98 as an efficient adsorbent: kinetics, isotherms, and thermodynamics study *J. Mater. Environ. Sci.* **8** 2993–3012
[97] Azimi A, Azari A, Rezakazemi M and Ansarpour M 2017 Removal of heavy metals from industrial wastewaters: a review *ChemBioEng Rev.* **4** 37–59
[98] Renu N A, Agarwal M and Singh K 2017 Methodologies for removal of heavy metal ions from wastewater: an overview *Interdiscip. Environ. Rev.* **18** 124
[99] Verma B and Balomajumder C 2020 Hexavalent chromium reduction from real electroplating wastewater by chemical precipitation *Bull. Chem. Soc. Ethiop.* **34** 67–74
[100] Gunatilake S K 2015 Methods of removing heavy metals from industrial wastewater *J. Multidiscip. Eng. Sci. Stud.* **1** 13–8
[101] Wang L K, Chen J P, Hung Y-T and Shammas N K (ed) 2011 *Membrane and Desalination Technologies* (Totowa, NJ: Humana)
[102] Rahmati N O, Pourafshari Chenar M and Azizi Namaghi H 2017 Removal of free active chlorine from synthetic wastewater by MEUF process using polyethersulfone/titania nanocomposite membrane *Sep. Purif. Technol.* **181** 213–22
[103] Huang J, Qi F, Zeng G, Shi L, Li X, Gu Y and Shi Y 2017 Repeating recovery and reuse of SDS micelles from MEUF retentate containing Cd^{2+} by acidification UF *Colloids Surfaces* A **520** 361–8
[104] Tanhaei B, Pourafshari Chenar M, Saghatoleslami N, Hesampour M, Laakso T, Kallioinen M, Sillanpää M and Mänttäri M 2014 Simultaneous removal of aniline and nickel from water by micellar-enhanced ultrafiltration with different molecular weight cut-off membranes *Sep. Purif. Technol.* **124** 26–35
[105] Fu F and Wang Q 2011 Removal of heavy metal ions from wastewaters: a review *J. Environ. Manage.* **92** 407–18
[106] Paulino A T, Minasse F A S, Guilherme M R, Reis A V, Muniz E C and Nozaki J 2006 Novel adsorbent based on silkworm chrysalides for removal of heavy metals from wastewaters *J. Colloid Interface Sci.* **301** 479–87
[107] Oyaro N, Juddy O, Murago E N M and Gitonga E 2007 The contents of Pb, Cu, Zn and Cd in meat in Nairobi, Kenya *J. Food Agric. Environ.* **5** 119–21
[108] Rivas B L, Pereira E and Maureira A 2009 Functional water-soluble polymers: polymer–metal ion removal and biocide properties *Polym. Int.* **58** 1093–114
[109] Baharuddin N H, Sulaiman N M N and Aroua M K 2015 Removal of heavy metal ions from mixed solutions via polymer-enhanced ultrafiltration using starch as a water-soluble biopolymer *Environ. Prog. Sustain. Energy* **34** 359–67
[110] Abdullah N, Yusof N, Lau W J, Jaafar J and Ismail A F 2019 Recent trends of heavy metal removal from water/wastewater by membrane technologies *J. Ind. Eng. Chem.* **76** 17–38
[111] Samaei S M, Gato-Trinidad S and Altaee A 2020 Performance evaluation of reverse osmosis process in the post-treatment of mining wastewaters: case study of Costerfield mining operations, Victoria, Australia *J. Water Process Eng.* **34** 101116

[112] Ahmed S F *et al* 2021 Recent developments in physical, biological, chemical, and hybrid treatment techniques for removing emerging contaminants from wastewater *J. Hazard. Mater.* **416** 125912

[113] Al-Amshawee S, Yunus M Y B M, Azoddein A A M, Hassell D G, Dakhil I H and Hasan H A 2020 Electrodialysis desalination for water and wastewater: a review *Chem. Eng. J.* **380** 122231

[114] Nemati M, Hosseini S M and Shabanian M 2017 Novel electrodialysis cation exchange membrane prepared by 2-acrylamido-2-methylpropane sulfonic acid; heavy metal ions removal *J. Hazard. Mater.* **337** 90–104

[115] Gherasim C V, Křivčík J and Mikulášek P 2014 Investigation of batch electrodialysis process for removal of lead ions from aqueous solutions *Chem. Eng. J.* **256** 324–34

[116] Min K J, Kim J H and Park K Y 2021 Characteristics of heavy metal separation and determination of limiting current density in a pilot-scale electrodialysis process for plating wastewater treatment *Sci. Total Environ.* **757** 143762

[117] Basha C A, Selvi S J, Ramasamy E and Chellammal S 2008 Removal of arsenic and sulphate from the copper smelting industrial effluent *Chem. Eng. J.* **141** 89–98

[118] Worch E 2021 *Adsorption Technology in Water Treatment: Fundamentals, Processes, and Modeling* (Berlin: De Gruyter)

[119] Yunus Z M, Al-Gheethi A, Othman N, Hamdan R and Ruslan N N 2020 Removal of heavy metals from mining effluents in tile and electroplating industries using honeydew peel activated carbon: a microstructure and techno-economic analysis *J. Clean. Prod.* **251** 119738

[120] Ao W *et al* 2018 Microwave assisted preparation of activated carbon from biomass: a review *Renew. Sustain. Energy Rev.* **92** 958–79

[121] Kosheleva R I, Mitropoulos A C and Kyzas G Z 2019 Synthesis of activated carbon from food waste *Environ. Chem. Lett.* **17** 429–38

[122] Rivera-Utrilla J, Sánchez-Polo M, Gómez-Serrano V, Álvarez P M, Alvim-Ferraz M C M and Dias J M 2011 Activated carbon modifications to enhance its water treatment applications. An overview *J. Hazard. Mater.* **187** 1–23

[123] Dutta S, Gupta B, Srivastava S K and Gupta A K 2021 Recent advances on the removal of dyes from wastewater using various adsorbents: a critical review *Mater. Adv.* **2** 4497–531

[124] Mollah M Y A, Morkovsky P, Gomes J A G, Kesmez M, Parga J and Cocke D L 2004 Fundamentals, present and future perspectives of electrocoagulation *J. Hazard. Mater.* **114** 199–210

[125] Kobya M, Can O T and Bayramoglu M 2003 Treatment of textile wastewaters by electrocoagulation using iron and aluminum electrodes *J. Hazard. Mater.* **100** 163–78

[126] Ghime D and Ghosh P 2020 Advanced oxidation processes: a powerful treatment option for the removal of recalcitrant organic compounds *Advanced Oxidation Processes—Applications, Trends, and Prospects* (Rijeka: InTech)

[127] Gutierrez-Mata A G, Velazquez-Martínez S, Álvarez-Gallegos A, Ahmadi M, Hernández-Pérez J A, Ghanbari F and Silva-Martínez S 2017 Recent overview of solar photocatalysis and solar photo-Fenton processes for wastewater treatment *Int. J. Photoenergy* **2017** 8528063

[128] Mahalingam S and Ahn Y H 2018 Improved visible light photocatalytic activity of rGO–Fe_3O_4–NiO hybrid nanocomposites synthesized by: *in situ* facile method for industrial wastewater treatment applications *New J. Chem.* **42** 4372–83

[129] Chen D *et al* 2020 Photocatalytic degradation of organic pollutants using TiO_2-based photocatalysts: a review *J. Clean. Prod.* **268** 121725

[130] Qiu X, Wan Z, Pu M, Xu X, Ye Y and Hu C 2022 Synthesis and photocatalytic activity of Pt-deposited TiO_2 nanotubes (TNT) for rhodamine B degradation *Front. Chem.* **10** 584

[131] Barrera-Salgado K E, Ramírez-Robledo G, Álvarez-Gallegos A, Pineda-Arellano C A, Sierra-Espinosa F Z, Hernández-Pérez J A and Silva-Martínez S 2016 Fenton process coupled to ultrasound and UV light irradiation for the oxidation of a model pollutant *J. Chem.* **2016** 4262530

[132] Ahmed Y, Lu J, Yuan Z, Bond P L and Guo J 2020 Efficient inactivation of antibiotic resistant bacteria and antibiotic resistance genes by photo-Fenton process under visible LED light and neutral pH *Water Res.* **179** 115878

[133] Ncube P, Pidou M, Stephenson T, Jefferson B and Jarvis P 2018 Consequences of pH change on wastewater depth filtration using a multimedia filter *Water Res.* **128** 111–9

[134] Taşar Ş, Kaya F and Özer A 2014 Biosorption of lead(II) ions from aqueous solution by peanut shells: equilibrium, thermodynamic and kinetic studies *J. Environ. Chem. Eng.* **2** 1018–26

[135] Es-Sahbany H, Berradi M, Nkhili S, Hsissou R, Allaoui M, Loutfi M, Bassir D, Belfaquir M and El Youbi M S 2019 Removal of heavy metals (nickel) contained in wastewater-models by the adsorption technique on natural clay *Mater. Today Proc.* **13** 866–75

[136] Chen H, Zhao J, Dai G, Wu J and Yan H 2010 Adsorption characteristics of Pb(II) from aqueous solution onto a natural biosorbent, fallen *Cinnamomum camphora* leaves *Desalination* **262** 174–82

[137] Xu X, Yang Y, Wang G, Zhang S, Cheng Z, Li T, Yang Z, Xian J, Yang Y and Zhou W 2020 Removal of heavy metals from industrial sludge with new plant-based washing agents *Chemosphere* **246** 125816

[138] Tirtom V N and Dinçer A 2021 Effective removal of heavy metals from an aqueous solution with poly(*N*-vinylimidazole-acrylamide) hydrogels *Sep. Sci. Technol.* **56** 912–24

[139] Moussavi G and Barikbin B 2010 Biosorption of chromium(VI) from industrial wastewater onto pistachio hull waste biomass *Chem. Eng. J.* **162** 893–900

[140] Alothman Z A, Bahkali A H, Khiyami M A, Alfadul S M, Wabaidur S M, Alam M and Alfarhan B Z 2020 Low cost biosorbents from fungi for heavy metals removal from wastewater *Sep. Sci. Technol.* **55** 1766–75

[141] Almomani F, Bhosale R, Khraisheh M, Kumar A and Almomani T 2020 Heavy metal ions removal from industrial wastewater using magnetic nanoparticles (MNP) *Appl. Surf. Sci.* **506** 144924

[142] Hu X, Hu Y, Xu G, Li M, Zhu Y, Jiang L, Tu Y, Zhu X, Xie X and Li A 2020 Green synthesis of a magnetic β-cyclodextrin polymer for rapid removal of organic micro-pollutants and heavy metals from dyeing wastewater *Environ. Res.* **180** 108796

[143] Musso T B, Parolo M E and Pettinari G 2019 pH, ionic strength, and ion competition effect on Cu(II) and Ni(II) sorption by a Na-bentonite used as liner material *Polish J. Environ. Stud.* **28** 2299–310

[144] Joseph L, Jun B M, Flora J R V, Park C M and Yoon Y 2019 Removal of heavy metals from water sources in the developing world using low-cost materials: a review *Chemosphere* **229** 142–59

[145] Villarín M C and Merel S 2020 Paradigm shifts and current challenges in wastewater management *J. Hazard. Mater.* **390** 122139

[146] Ma J, Li F, Qian T, Liu H, Liu W and Zhao D 2017 Natural organic matter resistant powder activated charcoal supported titanate nanotubes for adsorption of Pb(II) *Chem. Eng. J.* **315** 191–200
[147] Lin L, Xu X, Papelis C and Xu P 2017 Innovative use of drinking water treatment solids for heavy metals removal from desalination concentrate: synergistic effect of salts and natural organic matter *Chem. Eng. Res. Des.* **120** 231–9

Muhammad Ikram, Muhammad Ahsaan Bari and Junaid Haider

Chapter 2

Mechanisms and characteristics of photocatalysis

Photocatalysis is regarded as one of the most promising approaches for the degradation of industrial wastewater owing to the fact that it does not produce any secondary contaminants, is economical, and is favorable to the environment. The mechanism of photocatalysis results in larger pollutants, such as dyes, being broken down into smaller molecules that are less harmful, such as carbon dioxide and water. The conventional and advanced processes for the degradation of pollutants are both included in photocatalysis. The advanced type II mechanism, which is widely employed by researchers and produces efficient degradation outcomes, is one of the advanced processes. In addition, the Z-scheme is a developing method of pollutant control that demonstrates successful results. A redox-mediator-free Z-scheme photocatalytic system is the ideal choice for pollutant degradation, carbon dioxide and water splitting, and conversion because it can decrease the electron transfer distance and be used in liquid-phase and gas-phase conditions. The purpose of this chapter is to give a comprehensive examination of heterogeneous and homogeneous photocatalysts, including examples that best illustrate each kind. In addition to this, a brief overview will be given of the numerous parameters that govern the mechanism routes and processes of photocatalysis. The quick recombination of charge that occurs during heterogeneous photocatalysis is a key disadvantage that slows down the degradation process. Several different approaches, such as doping and heterostructuring, have been investigated in depth as potential solutions to this problem.

2.1 Photocatalysis

Natural photosynthesis was the motivation behind the invention of photocatalysis. The photocatalysis process is non-toxic and chemically stable and operates at room

doi:10.1088/978-0-7503-5517-9ch2

temperature. The word 'photo' is related to photons, while the word 'catalyst' describes a substance that alters the rate of a process whenever it is present. Together, these two terms make up the term 'photocatalyst'. Photocatalysts are substances that, when subjected to light, alter the speed at which a chemical process takes place. In photocatalysis, processes happen through the utilization of light in conjunction with a semiconductor. A photocatalyst is a type of catalyst that absorbs light and acts as a catalyst for various chemical processes. Semiconductors are the primary component of any photocatalyst. Conventionally, the term 'photocatalysis' refers to the process of harnessing the entire solar light spectrum to chemically break down big molecules into small, less toxic or lower in molecular weight species. It is a light-induced photocatalytic process, which effectively oxidizes or decreases organic pollutants through much simpler redox reactions [1]. It is a potential method for reducing energy consumption that converts light energy directly into chemical energy that can be utilized to break down and remove refractory organic compounds [2]. Due to the convenience, ability to operate at ambient temperatures and pressures, complete degradation of toxins, cost-effectiveness, lack of secondary pollution, and reduction of pollution levels to ppm and ppb limits, photocatalysis significantly outperforms other conventional approaches [3]. Additionally, photocatalysis is a sustainable option for environmental purification, as it eliminates the need for any conventional wastewater treatment techniques. A few distinctive benefits of photodegradation are enumerated in figure 2.1.

The photocatalysis process can be carried out in both an aqueous medium and in air [4]. Other organic contaminants besides dyes can be treated using the photocatalysis method, including antibiotics, synthetic chemicals, polymers, persistent organic pollutants, organic solvents, ionic liquids, surfactants, polycyclic aromatic

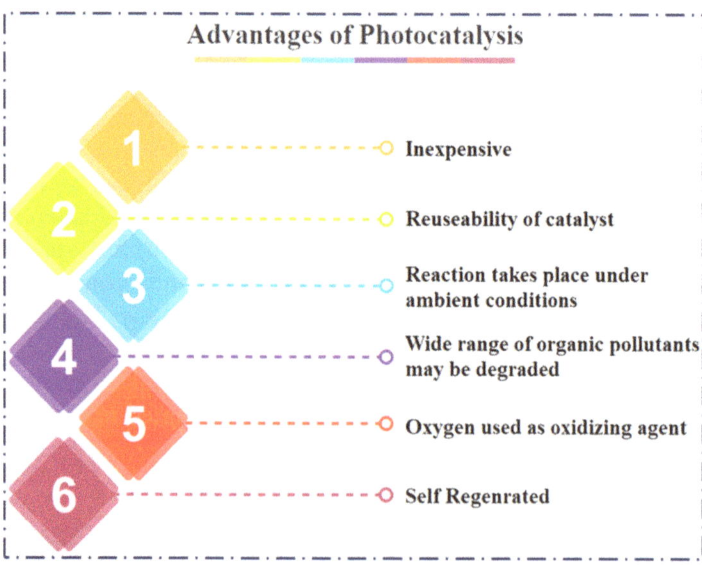

Figure 2.1. The principal benefits of photocatalysis for treating wastewater.

hydrocarbons [5], and so on. Additionally, this technique is applied to split water and reduce CO_2 in order to generate fuel and other valuable products [6]. Photocatalysis presents several challenges, the most significant of which are the difficulty in separating the required catalyst when it already exists as slurry and the need for ultraviolet light to activate the surface. Commonly applied photocatalysis methods include heterogeneous and homogeneous photocatalysis [7].

2.2 Homogeneous photocatalysis

Homogeneous photocatalysis is a particular kind of photocatalytic reaction that occurs when the reactant and the photocatalyst exist in the same state, whether that state is liquid, gas, or solid. Both homogeneous and heterogeneous conditions can be used to carry out photocatalytic degradation reactions; however, heterogeneous conditions are more commonly used. Solid semiconductors are used more extensively than soluble photocatalysts because they are less expensive, more robust, and easier to recover. Although heterogeneous photocatalysis has many advantages, homogeneous systems should not be overlooked; in fact, homogeneous systems may likely take on a more significant role in the near future [8]. Since molecules in their excited states are more potent oxidants (by accepting an electron at the vacant site in the HOMO) and reductants (by donating electrons in an empty orbital) than their ground states, photocatalysis may also take place throughout the homogeneous phase using molecules as catalysts. This results in an electron transfer process. Another possibility is that the structure of excited states favors the transfer of atoms. Dye molecules display intense banding not only in the ultraviolet spectrum but also in the visible spectrum during the homogeneous photocatalysis process. However, the usage of these compounds is not without its drawbacks, such as the difficulty in recovering them after a reaction, the photostability of these compounds, and the potential toxicity of these compounds. Even though exploratory and mechanistic investigations have been conducted, the application of homogeneous photocatalysis is still in the early stages. This is probably because of the reasons listed above. Homogeneous photocatalytic techniques, such as the Fenton and photo-Fenton procedures, have seen widespread application to destroy organic compounds found in contaminated streams. The following section will provide a concise mechanical explanation of the aforementioned procedures.

2.2.1 Fenton methods

In 1876 the scientific research of Fenton suggested that acid may be neutralized by combining H_2O_2 and Fe^{2+}, however, most people consider that Fenton chemistry did not begin until 1894, when he published a more in-depth investigation on the solid enhancement of oxidation of tartaric acid using such a reagent. The stoichiometry of the reaction that takes place between H_2O_2 and Fe^{2+} was investigated between the years 1901 and 1928. As a radical mechanism for the catalytic breakdown of H_2O_2 by iron salts, the outstanding practical applicability of Fenton's reagent for the oxidation of organic molecules was initially postulated in the 1930s [9].

The homogeneous Fenton concept is based on the production of reactive oxygen species (ROS). During the Fenton process, H_2O_2 decomposes when iron ions are present in aqueous solutions. The primary chemical stages are demonstrated by the following equations [10]:

$$Fe^{2+} + H_2O_2 \rightarrow Fe^{3+} + OH^- + OH \tag{2.1}$$

$$OH + H_2O_2 \rightarrow HO_2 + H_2O \tag{2.2}$$

$$Fe^{2+} + OH \rightarrow Fe^{3+} + OH^- \tag{2.3}$$

$$Fe^{3+} + HO_2 \rightarrow Fe^{2+} + O_2 + H^+ \tag{2.4}$$

$$OH + OH \rightarrow H_2O_2 \tag{2.5}$$

$$\text{Organic contaminants} + OH \rightarrow \text{Degradation products.} \tag{2.6}$$

Fenton-like processes involve the utilization of different metals at low oxidation states, including copper and cobalt [11]:

$$Cu^+ + H_2O_2 \rightarrow Cu^{2+} + OH^- + OH. \tag{2.7}$$

It is common knowledge that the treatment capabilities of Fenton's reagent (Fe^{2+}/H_2O_2) change based on the ratio of H_2O_2 to $FeSO_4$. The therapy has a greater tendency to produce the effect of chemical coagulation if the amount of Fe^{2+} used is greater than that of H_2O_2. When these two quantities are switched, the therapy has a greater propensity to have the impact of chemical oxidation. Figure 2.2 demonstrates the breakdown of rhodamine B dye using the Fenton technique. In addition to significant dosages of H_2O_2 and iron salts, the practical implementation of a homogeneous Fenton reaction also involves precision pH control because pH is a

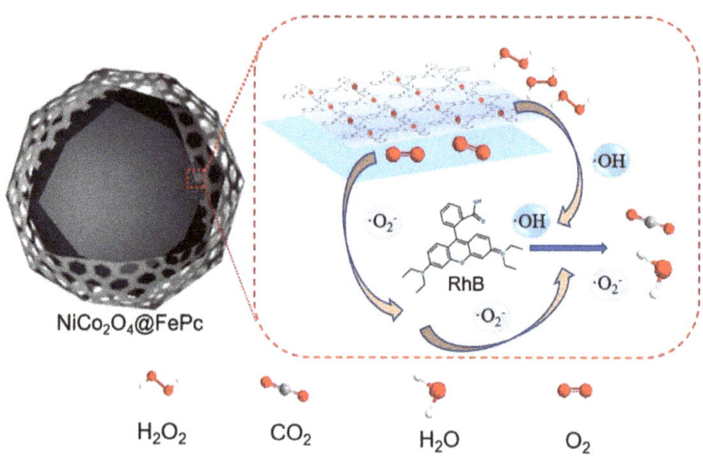

Figure 2.2. Schematic for the Fenton reaction triggered by the combination of $NiCo_2O_4$@FePc. (Reproduced with permission from [14]. Copyright 2020 Elsevier.)

key element that restricts the Fenton catalytic process in the existence of Fe (III) [12]. When the pH rises over 3, the Fe^{3+} starts to precipitate [13], forming a hydroxide that contains water and a relatively low activity level. In order to keep the environment at an acidic level during the reaction, it is necessary to introduce a significant quantity of acid, which is often sulfuric acid. After the reaction, the pH level must be returned to neutral before it may be released. The increased number of processes, in addition to the added expense, makes widespread implementation of this method unlikely.

In order to increase the effectiveness of the photocatalytic Fenton process, it is essential to find solutions to the following problems: how to facilitate the redox cycle of Fe (III)/Fe (II) and electron transfer, and how to achieve the most efficient use of solar radiation and effective utilization of H_2O_2. A substantial amount of research has been conducted, which has acted as both a source of direction and motivation for the continued growth of advanced photocatalytic oxidation.

2.2.2 Photo-Fenton method

The photo-Fenton reaction occurs when the material is exposed to light with a wavelength close to 600 nm, whereas the Fenton reaction does not involve any light irradiation. The former was not discovered until the 1960s, and it continues to be one of the widely utilized advanced oxidation processes as it breaks down high loadings of organic molecules even in very salty environments [15]. Hydroxyl radical production increases when hydrogen peroxide, ultraviolet light, and an ion with the charge states Fe^{2+} or Fe^{3+} are combined. As a result, the pace at which organic contaminants are degraded is sped up. The Fenton reaction causes an accumulation of Fe^{3+} ions and when all the Fe^{2+} ions have been consumed, the process can no longer proceed. In the photo-Fenton mechanism, ferrous ions are produced through the photochemical reproduction of ferrous ions through photoreduction of ferric ions, and the freshly formed ferrous ions interact once more with H_2O_2 to produce ferric ions and hydroxyl radicals, and the cycle continues:

$$Fe^{3+} + H_2O + h\nu \rightarrow Fe^{2+} + OH + H^+ \tag{2.8}$$

$$Fe^{3+} + H_2O_2 + h\nu \rightarrow Fe^{2+} + HO_2 + H^+. \tag{2.9}$$

Since the hydroxyl Fe^{3+} ions are efficiently soluble at a pH of 3.0, the photocatalytic performance of the $Fe(OH)^{2+}$ ion is enhanced at this pH. Therefore, the photo-Fenton process operates more efficiently. The use of sunshine rather than UV irradiation can cut costs in some instances; nevertheless, this results in a slower rate of breakdown of contaminants. According to the reports, acidic conditions, with a pH of around 3, are advantageous. As carbonic acid has poor reactivity with hydroxyl radicals, this may be predominantly attributed to the transformation of bicarbonate and carbonate species.

Due to the photo-Fenton procedures used to break down organic pollutants, carboxylic acids are produced, which react with iron to form complexes. To determine the contributions of oxalic acid, citric acid, formic acid, and malonic acid, a kinetic model mimicking the hydroxyl radical formation and iron redox cycle has been built. It has been discovered that oxalic and citric acid sped up the rate of both the photo-Fenton reaction as well as the iron redox cycle, but the effects of malonic acid and formic acid on the degradation rate were minor and actually negative. In contrast, it was found that carboxylic acids had no discernible impact on the Fenton-like reactions. These photo-Fenton reaction results have been explained by variations in the absorbance of UV light caused by the formed Fe^{3+} carboxylic acid complexes [16].

A few limitations of the homogeneous photo-Fenton method include the narrow pH range over which the reactions can occur, the need to extract the precipitated catalyst after the process, as well as the possibility for inactivation by specific ion complexing reagents, such as phosphate anions [17]. Homogeneous photo-Fenton also has a few advantages, such as the ability to perform the reaction at room temperature. There is also the possibility of utilizing heterogeneous solid Fenton nanocatalysts as an alternative technique. These nanocatalysts can be transition-metal-containing zeolites, clays, bentonites, etc. Because of their one-of-a-kind characteristics, including their ion exchange capacity, large surface area, and microporous structure, the utilization of synthetic zeolites shows great potential for the future [18].

2.3 Heterogeneous photocatalysis

Heterogeneous photocatalysis refers to photocatalytic reactions where the reactant and semiconductor are in distinct phases. Photocatalytic reactions can proceed through either a homogeneous or heterogeneous pathway; however, in recent years, heterogeneous photocatalysis has been the subject of a significantly greater amount of research. This is due to the fact that heterogeneous photocatalysis has the potential to be utilized in numerous applications related to the environment and energy and in organic compound synthesis. The photocatalytic reaction in heterogeneous photocatalysis necessitates the establishment of an interface between a solid photocatalyst (such as a semiconductor or metal) and another medium (such as a gas or fluid) that contains the products and reactants of the reaction. For this reason, the phrase ‘heterogeneous photocatalysis’ is often reserved for circumstances in which a light-absorbing photocatalyst is applied, and this photocatalyst is in contact with either a gas or a liquid state. Numerous scholars have devoted many good papers, reviews, and books to the diverse reactions of heterogeneous photocatalysis [19] and it has been the subject of ongoing investigation since its early stages. It has been demonstrated that heterogeneous photocatalysis is an efficient method for removing organic pollutants from the atmosphere and water [20]. These pollutants include dyes, insecticides, and poisonous chemical compounds. It does this by utilizing sunlight in conjunction with a semiconductor photocatalyst, which speeds up the process of cleaning up environmental toxins and highly harmful compounds [21].

2.3.1 The Langmuir–Hinshelwood model

Applying the Langmuir–Hinshelwood (L–H) process to heterogeneous photocatalysis, it is possible to comprehend the production of holes and electrons brought on by the catalyst's photoexcitation. Adsorption of the dye molecule onto the surface of the photocatalyst is the first step in the procedure. The next stage includes the movement of the photogenerated hole towards the surface, where they are captured by an adsorbed dye molecule and which causes them to form an unstable intermediate. This intermediate will be broken down further in the subsequent phase when it reacts with electrons. After that, the degradants are subsequently removed from the surface by a desorption process. After the process, the dye is transformed into a form less toxic for public health, and the catalyst's surface is returned to its initial state [22]. The following equations show the complete reaction mechanism:

$$\mathrm{S} + \mathrm{Dye} \rightarrow \mathrm{S} \cdot \mathrm{Dye} \tag{2.10}$$

$$\text{Photocatalyst} + h\nu \rightarrow \mathrm{h}^{+} + \mathrm{e}^{-}\ \text{(photogeneration of excitons)} \tag{2.11}$$

$$\mathrm{S} \cdot \mathrm{Dye} + \mathrm{h}^{+} \rightarrow \text{unstable intermediates} \tag{2.12}$$

$$\text{Unstable intermediates} + \mathrm{e}^{-} \rightarrow \mathrm{S} \cdot \text{Products} \rightarrow \mathrm{S} + \text{products.} \tag{2.13}$$

The amount of dye molecules that bind to the catalyst directly relates to the efficiency of photocatalysis and is influenced by the amount of dye in the solution. The L–H expression can be written out in its more straightforward version as [23]

$$1/r = 1/k_{\mathrm{r}} + 1/(k_{\mathrm{r}}k_{\mathrm{a}}C), \tag{2.14}$$

where r is the rate of oxidation of the reactant (mg l^{-1} min^{-1}), k_{r} is the specific rate of a reaction constant (mg l^{-1} min^{-1}), k_{a} is the reactant equilibrium constant (mg^{-1}), and C is the concentration of dye.

2.3.2 The Eley–Rideal model

In the year 1938 D D Eley and E K Rideal invented this particular kinetic model. In accordance with this paradigm, one of the reactants only becomes entangled on the photocatalyst's surface, whereas the other reactant remains accessible in the mobile phase. In their approach, the holes remain anchored to the surface while the dye is left free to float in the solution. When the holes reach the surface of the catalyst, some defects (D) on the surface cause them to become stuck. This results in the formation of the surface-active center. Following this, the active centers combine with the dye molecules to produce an unstable species. This species is then degraded, and the surface defect is returned to its initial shape. Some active centers could also break down when combined with electrons [22].

2.4 Various mechanisms of photocatalysis

2.4.1 Conventional mechanism

The excitation of electrons from the valance band (VB) of the nanomaterials to the conduction band (CB) by photons with photon energy higher than the energy gap of the semiconductor results in the formation of photogenerated holes (h^{+}) in the VB.

These photoexcited electron and hole couples might either recombine to generate thermal energy, which would stop the photocatalytic activity, or diffuse to the photocatalyst's surface and react with the adsorbed molecules. An oxidative potential (+2.8 V) is produced when the photogenerated holes (h^+) react with the water, producing hydroxyl radicals ($OH^{\cdot}$), whereas superoxide anion radical ($O_2^{-}\cdot$) is produced when the oxygen molecule traps the conduction band electrons. Such reactive radicals break down contaminants in a nonselective and speedy manner into species that are less likely to be affected [24]. The case study of ZnF nanoparticles (NPs) utilizing MB dye can be used to understand the mechanism of dye when it is present in the presence of semiconductor photocatalysts while the environment is illuminated by light. Because of the total breakdown of the methylene blue dye subjected to UV irradiation, figure 2.3(d) demonstrates that the blue dye was no longer present [25]. The following equations can be used to depict the potential interfacial reactions:

$$\text{photocatalyst} + h\nu(\text{UV}) \rightarrow e^-_{CB} + h^+_{VB} \tag{2.15}$$

$$O_2 + e^- (\text{CB}) \rightarrow O_2^- (\text{ads}) \tag{2.16}$$

$$H_2O (\text{ads}) + h^+ (\text{VB}) \rightarrow OH (\text{ads}) + H^+(\text{ads}) \tag{2.17}$$

$$\text{Dye} + h^+ (\text{VB}) \rightarrow \text{oxidation product} \tag{2.18}$$

$$\text{Dye} + e^- (\text{CB}) \rightarrow \text{degradation product.} \tag{2.19}$$

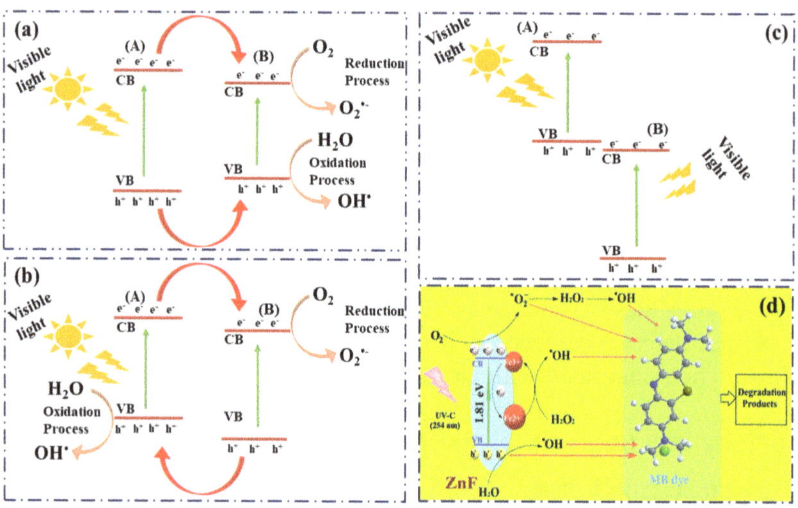

Figure 2.3. Schematic analysis of the (a) type I, (b) type II, and (c) type III mechanisms, and (d) the photocatalytic degradation process of MB dye onto ZnF photocatalyst. ((d) Reproduced with permission from [25]. Creative Commons Attribution 4.0 International License http://creativecommons.org/licenses/by/4.0/.)

2.4.2 Advanced mechanism

2.4.2.1 Type I/straddling gap mechanism

Two semiconductors make up a type I heterojunction; semiconductor A's valance band is lower than semiconductor B's, and semiconductor A's conduction band is higher than semiconductor B's; as a result, electrons and holes would move from semiconductor A's conduction band and valence band to semiconductor B's conduction band and valence band, respectively. Therefore, the associated processes will use semiconductor B as the photocatalytic active sites. For instance, Ma *et al* developed a type I heterojunction (Mo_2C/In_2S_3). This allows the photogenerated electrons from the In_2S_3 valence band to be translated to Mo_2C. This also allows the photogenerated holes of In_2S_3 to be migrated to the Mo_2C valence band. Finally, the migrating electrons can efficiently produce hydrogen [26]. For instance, in the case of a $ZnO/Fe_3O_4/g\text{-}C_3N_4$-50% nanocomposite, it is also hypothesized that the charge transfers to the surface and interacts immediately with methyl orange dye [27].

2.4.2.2 Type II/staggered gap mechanism

The semiconductor B possesses relatively low conduction and valence band positions in the type II heterojunction, which is mainly composed of a staggered band gap type. The photogenerated electrons at the conduction band of semiconductor A will move into the conduction band of semiconductor B, whereas the holes at the valence band of semiconductor B will move but also accumulate into the valence band of semiconductor A. As a result, the photocatalytic centers for photo redox reactions will be the conduction band at semiconductor B and the valence band at semiconductor A, offering such a heterojunction an effective structure enabling charge separation as well as collection. The type II heterojunction photocatalyst is the most straightforward technique to increase photocatalytic performance thanks to its effectiveness in segregating the photoexcited e^-/h^+ pairs through the band alignment over two semiconductors [28]. It should be emphasized that p–n junctions, where a space-charge zone occurs at the interface, would form if the semiconductors A and B are p- and n-type semiconductors. The electrical potential established by the diffusion of holes and electrons enhances the flow of holes and electrons into the conduction and valence band. Type II p–n semiconductor junctions will have more advantages than type I p–n junctions, such as more effective charge separation, longer charge carrier lifetimes, and isolated places for oxidization and reduction reactions. For instance, both materials in the $VO_2/CuWO_4$ type II n–n heterojunction are n-type semiconductors, substantially isolating the electron–hole pairs and minimizing the amount of recombination. VO_2 emits holes and electrons when exposed to visible light because it has a narrow band gap. This occurs as a result of the photoexcitation of VO_2. Electrons that have been photoexcited can quickly travel from the higher-lying conduction band of VO_2 to the lower-lying conduction band of $CuWO_4$ due to an internal electric field. Reactive species $OH^\cdot$ and $O_2^{\bullet -}$ are produced when photoexcited electron and hole molecules react with molecules of H_2O and O_2 that have been adsorbed [29].

2.4.2.3 Type III/broken-gap mechanism

The band positions of the type III heterojunction are more extreme than those of the type II heterojunction. This means that the bandgaps of semiconductors A and B do not overlap, which results in a broken-gap structure [30]. Semiconductor B has a lower conduction and valence band than semiconductor A in type III. Nevertheless, in type III alignment, neither electrons nor holes from one semiconductor may transfer to another. They can therefore be considered to be two different semiconductors as there is no heterojunction between them [31]. Figures 2.3(a)–(c) present a comparative analysis of type I, type II, and type III mechanisms.

2.4.2.4 Z-scheme mechanism

The direct Z-scheme heterojunction is characterized by the spatial flow of electrons from the conduction band of semiconductor B to the valence band of semiconductor A, and the drawing of the pathway resembles the letter 'Z'. For the Z-scheme electron transfer, the semiconductor A–B contact interface's energy level must be almost continuous, like a conductor. A Z-scheme heterojunction can occur under several conditions. The construction of the Z-scheme heterojunction for the direct contact of the semiconductors A and B is based on the Fermi energy level's rational realignment, which results from the redistribution of charges (figure 2.4(a)). This results in forming an internal electric field that speeds up the recombination of photoexcited holes in the valence band of semiconductor A and photoexcited electrons in the conduction band of semiconductor B. As a result, photoexcited holes in the valence band of semiconductor B and photoexcited electrons in the conduction band of semiconductor A can be accumulated and maintained [32–34]. The photogenerated holes at the valence band of semiconductor A and the photogenerated electrons at the conduction band of semiconductor B would transfer through the conductor and recombine, which would result in a high density of holes at the valence band of semiconductor B and electrons at the conduction band of semiconductor A for enhanced redox reactions in semiconductors that are contact-mediated with a conductor, such as carbon or metal-based material. In addition, the electron acceptor and donor pair can serve as an electron mediator for the Z-scheme junction building of two semiconductors that are not in direct contact with one another (figure 2.4(b)) [35].

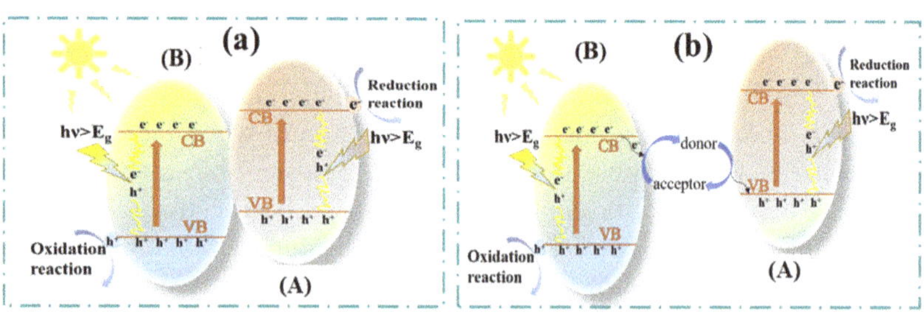

Figure 2.4. Schematic illustration of (a) direct Z-scheme charge transfer and (b) redox-mediator Z-scheme charge transfer. (Reproduced with permission from [38]. Copyright 2019 Elsevier.)

As recently revealed by Kim and Yong [36], those heterojunction types are occasionally not exclusively prevalent in a photocatalytic process containing two or three semiconductors; they can also be a combination or transition from one type to another [37]. They used straightforward boron (B) doping of C_3N_4/ZnO to show how the charge transfer mechanism changes from the Z-scheme to type II. When C_3N_4 is doped with low-electronegativity boron, its Fermi level rises by 0.4 V, surpassing that of ZnO. The outcome is a reversed band bending direction at the C_3N_4/ZnO junction due to the Fermi level alignment of B-doped C_3N_4 with ZnO. When type II B-doped C_3N_4/ZnO is compared to type II undoped C_3N_4/ZnO, the photocatalytic H_2 evolution rate increases steadily and dramatically by about 2.9 times (Z-scheme).

The development of a direct Z-scheme photocatalytic device offers enormous promise for alleviating energy shortages and environmental pollution. There are still many challenges to be solved, such as how to increase each semiconductor's light conversion efficiency and long-term stability as well as how to improve the contact interface between two parts to boost photocatalytic performance. A single-component semiconductor photocatalyst finds it challenging to overcome the significant difficulties described above when compared to the artificial Z-scheme photocatalytic system. The best option for pollutant degradation, CO_2, and water splitting conversion is a redox-mediator-free Z-scheme photocatalytic system that can minimize electron transfer distance and be used in liquid-phase and gas-phase environments. This is because the photocatalytic performance of redox-mediator Z-scheme photocatalytic systems is restricted by backward reactions, the reaction environment, long-distance electron transfer, and light-shielding effects.

2.5 Effects of operational parameters on photocatalysis

The efficiency of a photocatalyst in a certain condition can be thought of as the rate at which it can mineralize dye molecules in that condition. The effectiveness of photocatalysts can be altered by various conditions, some external while others are intrinsic to the system. The term 'internal factor' refers to variables within the photocatalyst, such as its physical structure and chemical make-up. When we talk about 'external factors', we refer to things such as pH, temperature, irradiation intensity, the photocatalyst concentration, contaminants in the reaction mixture, and the presence of additives.

2.5.1 Effect of pH

The starting pH of the solution is one of the most critical variables in photocatalytic reactions. This parameter controls the catalyst's surface electrical charge properties and the catalyst surface's ionization state, which impact the way dye compounds degrade [39]. Typically, the type of catalyst being employed will have an impact on the photocatalytic process's activity. Due to the numerous roles that pH plays, it is challenging to determine how the pH value affects the photodegradation process. The response of the catalyst and the pollutant to pH variations may be the cause of all of these things. The pH value impacts the compounds' ability to dissociate, the distribution of charges on the catalyst's surface, and the oxidation potential of the catalyst's valence band [40].

The point of zero charge pH refers to a characteristic of solid particles in aqueous samples (pH_{PZC}). The catalyst's properties, such as its surface charge, determine its PZC value. For instance, an ion on the catalyst's particle's surface can characterize its charge [41]. In the case of the negative charge particle, the other cations would be less adsorbed because hydrogen ions (protons, H^+) are adsorbed at a lower pH than other cations. On the other hand, as the concentration of hydroxide ions increases, anions will be adsorbed less if the surface is positively charged and the pH is elevated [42].

According to Zhu *et al* [43], the relationship between the pH effect and the photocatalysts' surface-charge characteristics can be described by the idea of a zero charge point (PZC). The point at which TiO_2 particles have no charge has been discovered to be pH 6.8. When the pH is lower than pH_{PZC} (pH 6.8) or the solution is acidic, the catalyst surface becomes positively charged, or vice versa [44]. This is shown by the following equations:

$$TiOH_{(Surface)} + H^+ \rightarrow TiOH_{2(surface)}{}^+ \tag{2.20}$$

$$TiOH_{(Surface)} + OH^- \rightarrow TiO_{(Surface)}{}^- + H_2O. \tag{2.21}$$

The adsorbent's surface becomes positively charged whenever the pH value is lower than the PZC value, and this causes the surface to become anion-attracting while simultaneously cation-repelling. In contrast, the surface becomes negative above PZC, as a result, cations are attracted to it while anions are repelled by it.

Additionally, pH can impact the entire photocatalytic dye degradation pathway and, therefore, the degradation rate. For instance, the principal oxidation species produced when the pH is lower is called positive holes; however, hydroxyl radical ($OH^\cdot$) production is solely accountable for oxidation whenever the pH is neutral or higher. These radicals can be generated in an alkaline solution, but electrostatic repulsion may prevent some areas on the surface of the semiconductor from absorbing dye anions, limiting the radicals' ability to degrade, and vice versa [45]. It is proposed that $OH^\cdot$ radicals are essential for the –N=N conjugated structure in azo dyes to fission during a TiO_2-aided photocatalytic degradation. Electrophilic assault by hydroxyl radical is the primary means of breaking the azo bond (–N=N–). According to Riga *et al*, the dye degradation and decolorization rate accelerate with rising pH [46]. On the other hand, lower pH results in an excess of H^+ ions, and all these H^+ ions interface with azo linkage, lowering the electron densities of the azo group. As a result, an electrophilic process reduces the hydroxyl reactivity of $OH^\cdot$.

Hanafi and colleagues [47] explored how the removal of the Remazol Brilliant Blue (RBB) dye was affected by pH 3, 5, 7, 9, and 11. The researchers used $EGZrO_2$ as the catalyst. According to the results, when the pH value was raised, the proportion of degraded RBB dropped significantly. When reactions occur on the semiconductor surface, the metal oxide semiconductor typically displays an amphoteric behavior. This implies that the behavior will affect the catalyst's surface-charge properties [48]. Therefore, more negatively charged anions of RBB are present on the surface of the photocatalyst whenever the solution's pH value is lower. This is because naturally occurring positive charge anions are present at pH levels lower than pH_{PZC}. This phenomenon is primarily attributed to the increase in the electrostatic force between the RBB dye molecule and the catalyst, which will improve the dye's ability to degrade at low pH levels.

2.5.2 Dosage of semiconductor photocatalyst

When the dose of a catalyst is increased, the photocatalytic degradation usually increases at the same time. This is because there is a larger surface area accessible for adsorption, and an irradiated photocatalyst generates a greater amount of hydroxyl radicals. When there is less catalyst present in the reactor, more light is allowed to get through, but the amount of radiation utilized in the reaction is reduced. However, the degradation rate may be slowed down due to an increase in the catalyst suspension that is more than the optimal catalyst loading. As a result, fewer photocatalysts are activated and photons have a shallower light penetration depth [49].

Nguyen *et al* [50] recorded the impact of Pd-doped TiO_2 photocatalyst dosage (between 0.5 and 2.0 g l^{-1}) on the breakdown of methyl orange (MO) as well as methyl benzoate (MB) at a neutral pH. The findings led researchers to conclude that an increase in the number of oxidative hydroxyl radicals produced at the catalyst surface causes an appreciable rise in the apparent rate constant from 0.5 to 1.0 g l^{-1}. The degradation rate for MO and MB decreased significantly when the dosage was increased to 2.0 g l^{-1}, but the difference was minor for MO degradation when it increased to 1.5 g l^{-1}. The increased opacity of the solution as a result of the additional catalyst loading might be detrimental to the penetration of light.

2.5.3 Concentration of dyes

Following the Beer–Lambert rule, increasing the initial concentration of the dye causes a reduction in the path length of photons that are allowed to enter the solution. This causes a drop in the number of photons that are absorbed on the photocatalyst surface, which, in turn, causes a decrease in the rate at which photocatalytic breakdown occurs [51]. Increasing the initial dye concentration causes a proportional rise in the photodegradation (PD) rate up to a certain point, after which it levels off and then begins to fall when the concentration is increased beyond the limit. This is probably happening because there is a growing quantity of dye, which results in a greater amount of absorption on the photocatalyst surface. Another probable cause for decreased efficiency in degradation is that it is difficult for light to penetrate the surface of the photocatalyst [52]. Furthermore, the dyes adsorbed on the photocatalyst's surface prevent the interaction of the adsorbed molecules with the generated reactive radicals [53]. Research using unsupported multi-walled carbon nanotubes (MWCNTs) and supported Mn–Co NPs photocatalysts with PD malachite green dye in an aqueous medium showed that dye degradation rate reduced as the initial concentration of dye increased [54]. As MO dye concentration increases up to 10.0 mg l^{-1} during the degradation process utilizing Mg-doped TiO_2 and visible light irradiation, PD increases and subsequently falls as dye concentration increases [55].

2.5.4 The influence of additives and impurities

Various ions and chemicals are present in effluents, which are vital in controlling dye breakdown rates [56]. In most cases, wastewater contains ions that include NO_3^-, Ca^{2+}, NH_4^+, Mg^{2+}, Na^+, Al^{3+}, K^+, SO_4^{2-}, Cl^-, HCO_3^-, PO_4^{3-}, and Fe^{2+} [57–59]. The nature of many of these compounds is such that they can scavenge hydroxyl radicals and electrons. This means that they rapidly react with radicals, which drastically limits the number of $OH^·$ that have the potential to interact with molecules of the dye. As a direct consequence of this, the rate of the response is significantly slowed down. The pace of photocatalytic activity of Reactive Yellow 145 by TiO_2 was researched by Alahiane and colleagues to determine how the presence of a range of inorganic ions affected the reaction rate [57]. After this, it was discovered that the existence of Cl^- and SO_4^{2-} ions might increase the PD rate to a greater extent. Cl^- and SO_4^{2-} combine with the $OH^·$ radicals to produce the powerful oxidizing agents $SO_4^{·-}$ and $Cl^·$, which can quickly degrade dye molecules. Because HPO_4^{2-} and HCO_3^- adsorb onto the active centers present on the surface competitively, the dye molecules are prevented from being adsorbed. In addition, they play the role of hydroxyl scavengers, which significantly slows down the process. Due to the reaction between CH_3COO^- ions and the holes, methyl radicals and carbon dioxide are produced. As a result, the number of holes that are accessible to form hydroxyl radicals is decreased. The two researchers, Sudrajat and Babel, looked into the effect that sodium chloride (NaCl) salts have on the breakdown of amaranth dye [60]. It was discovered that the efficiency of degradation rose from 90.9% to 93.1% when the content of NaCl was increased from 0.3 to 1.2 g l^{-1}. However, when the dosage was increased to 9.6 g l^{-1}, the effectiveness of the degrading process decreased.

In the latter investigation, the initial improvement in effectiveness was credited to the photocatalyst's ability to react with Cl^- ions to produce $Cl^·$ ions. Because of their high oxidation potential (2.47 V), these $Cl^·$ ions can quickly break down the molecules of the dye. In addition, the $Cl^·$ ions interact with the holes, inhibiting recombination with the electrons. Nevertheless, when the concentration of Cl^- molecules continues to rise, they compete with the molecules of dye for the opportunity to react with the catalyst surface, resulting in significant slowing of the reaction rate. Additionally, research indicates that the carbonate radicals ($CO_3^{·-}$) produced by the reaction of carbonate ions with hydroxyl ions have a strong oxidizing potential and can destroy various compounds [61]. Additionally, they can interact with the molecules of dye found in most of the solution since they are more mobile and stable than hydroxyl radicals. Additionally, they can function as quenchers or aid in charge separation, which will stop recombination [61]. Similarly, bicarbonate ions have also accelerated photocatalysis at low concentrations [62].

2.5.5 Effect of temperature and irradiation intensity

2.5.5.1 Temperature

In addition to starting dye concentration and light intensity meeting the criteria for optimum settings, there is also a temperature range that satisfies the standards for perfect PD conditions. When the temperature is low, a spontaneous exothermic process known as the adsorption of the reactant is more likely to occur. However, when the temperature approaches 0 °C the apparent activation energy will rise. Additionally, despite the disadvantage of more active sites, a low temperature makes it easier to adsorb the final product. As a result, the reaction slows down and serves as the rate-limiting step at low reaction temperature since product desorption is slower than PD and reactant adsorption. In comparison, as the temperature approaches the boiling point of the solvent, which is typically water, the exothermic adsorption of reactants is no longer preferred. This ultimately limits the photodegradation process [63]. Electron–hole recombination is strongly encouraged when the reaction temperature exceeds 80 °C [64]. Because of their increased kinetic energy, dye molecules could be able to break free from the photocatalyst surface when the temperature is raised over a certain point [65], which would result in a reduction in the PD effectiveness. Because of this, the adsorption of dye molecules turns into the limiting step when the temperature is high. Consequently, the optimal reaction temperature for the efficient PD of dye molecules is considered to be somewhere between 20 °C and 80 °C [66].

The effect of temperature increase on Fe-doped ZnO-mediated PD of azophloxin dye was researched by Priyanka and Srivastava for the range of 30 °C–60 °C, and it was observed that the effectiveness of photocatalysis was enhanced greatly with the rise in temperature up to 50 °C [67]. However, it was observed that the effectiveness of the reaction dropped significantly when it was carried out at 60 °C. This occurred because of the effect of the high temperatures on the hydroxyl radicals, which caused them to self-scavenge.

2.5.5.2 Irradiation intensity

There is no correlation between the band gap sensitization process and the photocatalytic degradation rate or mechanism [68]. Despite this, extensive studies have shown that the photocatalytic degradation rate rises directly in accordance with the amount of light inside the environment [69]. Collazo *et al* examined the capacity of N-doped TiO_2 to degrade Direct Black 38 dye when subjected to a visible light source as well as natural sunshine [70]. It was established that nitrogen-doped TiO_2, which has a higher absorption in the visible area, exhibited a greater degree of dye degradation (60% in 6 h) when exposed to visible light. N-doped TiO_2 showed a degradation efficiency marginally greater than non-doped TiO_2 when subjected to sunlight irradiation.

2.5.6 Effect of oxidizing agents

During the photocatalytic activity, it is assumed that numerous oxygen species and reactive radicals are the key factors in the degradation of hazardous dyes [71]. ROS are produced by oxidants, improving the degradation efficiency and decreasing the required time [72]. The production of additional $OH^{\bullet}$ and the capacity of H_2O_2 to trap electrons contribute to an increase in PD effectiveness [73]. The reaction time of dye degradation is significantly reduced by adding H_2O_2. For instance, Rahmat *et al*

[74] degraded crystal violet dye over a long period of time, 280 min, but by adding H_2O_2 to the process they were able to cut the time needed for effective degradation down to 60 min. Nevertheless, excessive H_2O_2 produces less potent $^{\bullet}O_2H$ (hydroperoxyl) radicals [75]. The oxidation potential of these radicals is dramatically lower than $^{\bullet}OH$ radicals, despite being effective oxidants [76]. When present in more significant concentrations, H_2O_2 has the potential to recombine and scavenge with $^{\bullet}OH$, hence reducing the availability of these two molecules for the dye breakdown [77]. In the presence of hydrogen peroxide, azo dyes should deteriorate at a rate that is significantly quicker than normal [78]. It was determined that the rate of percent deterioration accelerates up to a critical point when the initial H_2O_2 concentration is raised, but after that the rate slows down [79].

2.5.7 Reusability of photocatalysts

Photocatalysts' capacity to be recovered and reused is crucial in assessing whether a catalyst can be considered sustainable. After several cycles, a material's photocatalytic activity was discovered to have slightly decreased [80]. The deposition of photosensitive hydroxides on a photocatalyst causes active sites on the photocatalyst to become blocked, which leads to a drop in photocatalytic effectiveness [81]. The integration of the magnetic and photocatalytic properties of NPs into the same framework is a notable method for achieving an effective separable catalytic system [82]. Magnetic catalysts exhibited a high photocatalytic activity when exposed to UV or visible light [83]. Additionally, they make it simple to isolate the degraded product from an aqueous solution following treatment with a magnetic field.

2.6 Methods for improving photocatalytic degradation efficiency

A photocatalyst's photocatalytic activity can be significantly influenced by several factors. Some of these parameters include the crystal's composition and its physical structure. It should be able to capture visible light, have a high charge separation efficiency, and have low charge recombination. Additionally, it should have a low energy band gap to make it simple to activate with photons. These are the characteristics that would make it an excellent photocatalyst. Since the band gap energies of various materials, such as TiO_2, are relatively high, these materials can only be stimulated by ultraviolet rays; visible light cannot do the job. Additionally, these materials have a high recombination rate attributed to the poor removal efficiency between the photoexcited electron–hole pairs. Doping, heterostructure, and shape manipulation are some approaches developed in recent years to help overcome these restrictions.

2.6.1 Doping schemes

Doping research for photocatalysts has been ongoing for a significant number of years. Doping is usually accomplished to increase a semiconductor's photocatalytic activity at lower energy levels, which can involve either non-metals or metals. Doping technically refers to adding foreign substances to the parent photocatalyst without causing the development of new structures, phases, or crystallographic shapes. Doping is done to increase the net separation of photogenerated charges and, as a consequence, efficiently harnessing the broad visible light component of the solar spectrum, which accounts for approximately 43% of the total, as opposed to the narrow ultraviolet component, which accounts for 5%. Doping is a technical term that refers to the process of introducing non-native components into the parent photocatalyst. As a direct consequence of this development, studying this specific facet of photocatalysis is becoming an increasingly active field.

2.6.1.1 Metal doping

In order to increase the photocatalytic efficacy of principal photocatalysts whenever subjected to visible light, second-generation visible-light-active photocatalysts (TiO_2 and ZnO) have also been doped with metals. This was achieved by introducing a metal ion into the semiconductor's crystalline structure. Most of the time, adding metal ions to a semiconductor can develop new energy levels between its valence and conduction bands. As a result, either ZnO or TiO_2's band gap energy decreases, changing the absorption properties in the visual spectrum. Because metals are far more active than other elements, electrons are rapidly excited, and the corresponding energy level structure of TiO_2 may be altered by metal ions doped with TiO_2. As a result, a TiO_2 system has a broader absorption spectrum [84, 85]. Metals have the ability, as seen in figure 2.5, to snare the electrons that are released due to TiO_2 excitation, and it is difficult for these electrons to return to their initial state. Upon metal doping the potential exists for metal ions to act as carrier-trapping centers. Compared to titanium ions, metal ions having valences greater than tetravalent have a higher likelihood of gaining electrons in this state, while metal ions having valences less than tetravalent operate as carrier traps. Ion doping can then block electron–hole pair recombination, which permits TiO_2 to create more holes and electrons than it would have otherwise been able to. Consequently, there is a increase in the

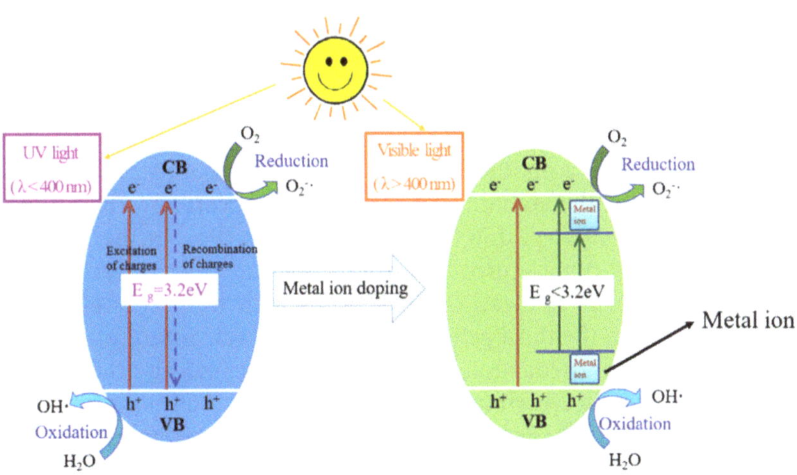

Figure 2.5. The mechanism of metal ion doping TiO_2. (Adapted from [95]. Creative Commons Attribution 4.0 International License http://creativecommons.org/licenses/by/4.0/.)

photocatalytic effectiveness of TiO_2 [86, 87]. In metals with a large number of valence states, electrons in the d orbitals have the potential to transition and enter the TiO_2 lattice. As a result, both the energy gap and the required energy for electron transitions are lowered [88, 89]. It is necessary to do this to boost the photocatalytic degradation capability of TiO_2 [90]. The energy level structure and crystalline morphology of nano-sized TiO_2 can be changed when transition group metal ions are doped with TiO_2 [91]. As a result, impurity energy levels are developed. Since photons with lower energy can also undergo transitions, this broadens the range of wavelengths over which they can absorb light as well as boosts the utility of visible light [92]. In the synthesis of visible active photocatalysts of the second generation, several metals have been explored. These metals include Fe, Cu, Co, Mn, and Ni for ZnO, and Ce, Er, La, Mo, and Cr for TiO_2. This type of doping has several disadvantages, however, its main downside is that metal ions may act as recombination sites for electron–hole pairs, which would reduce the photocatalytic degradation activity [93, 94].

Using the sol–gel process, Crisan *et al* [96] synthesized iron-doped TiO_2 and expanded the absorption spectrum to 546 nm. In addition, the iron content (w) of 2% resulted in a bandwidth of 3.03 eV when measured at a wavelength of 410 nm. The rate of nitrobenzene degradation was raised from 70% to 88% when 0.5% iron–TiO_2 was used compared to pure TiO_2. Whenever the energy bandwidth was decreased to 3.01 eV, Gnanasekaran and colleagues [97] discovered that the spectral absorption range of cobalt-doped TiO_2 increased (382 to 411 nm). After 240 min of treatment with a visible light catalytic agent, the rate of MO degradation reached 53%, which was favorable to the photocatalytic performance of the substance. A simple hydrothermal approach was used by Wang *et al* [98] to manufacture a Fe^{3+}-doped TiO^2 nanotube array catalyst. This resulted in an approximately 20% increase in the rate of MB's breakdown. The research also discovered that Fe^{3+} doping produced lattice traps TiO_2, which significantly increased the separation effect of electron–hole pairs.

2.6.1.2 Non-metal doping

Numerous studies have been conducted on TiO_2 and ZnO that have been doped with non-metal elements. These photocatalysts are considered to be the third generation of visible active photocatalysts. Doping photocatalysts with non-metals can significantly extend the amount of visible light they can absorb and reduce the amount of charge recombined due to photogeneration [99]. In the semiconductor lattice, the goal of the doping process using non-metal (mostly carbon, nitrogen, phosphorus, or sulfur) is to swap oxygen atoms with this other element. When exposed to visible light, the third generation of visible, active catalysts exhibits high levels of activity due to a reduction in their band gap caused by the mixing of the orbitals of the anions with 2p orbital of oxygen [100, 101]. Additionally, the presence of these elements inside the crystalline structure of the semiconductor leads to the production of certain defects, which slows down the recombination of a photogenerated pair [102, 103]. Using a form of computation known as 'first principles', Asahi *et al* reported on a potential approach of integrating N into the TiO_2 lattice [104].

The suggested approach includes substitutional NO-doping, and substitutional N-doping, interstitial NO-doping, and interstitial N-doping. In a study that was quite similar, Dunnill *et al* demonstrated that the incorporation of N into the TiO_2 lattice takes place through both substitutional and interstitial mechanisms (figure 2.6(c)) [105]. According to what they found, substitutional doping includes the replacement of oxygen, whereas interstitial doping includes the inclusion of N into the TiO_2 structure [105]. Although substitutional doping can narrow the band gap, it does so to a lower amount than interstitial doping can (which narrows the gap by 2.46 eV). The effectiveness of TiO_2 as a photocatalyst has been found to be

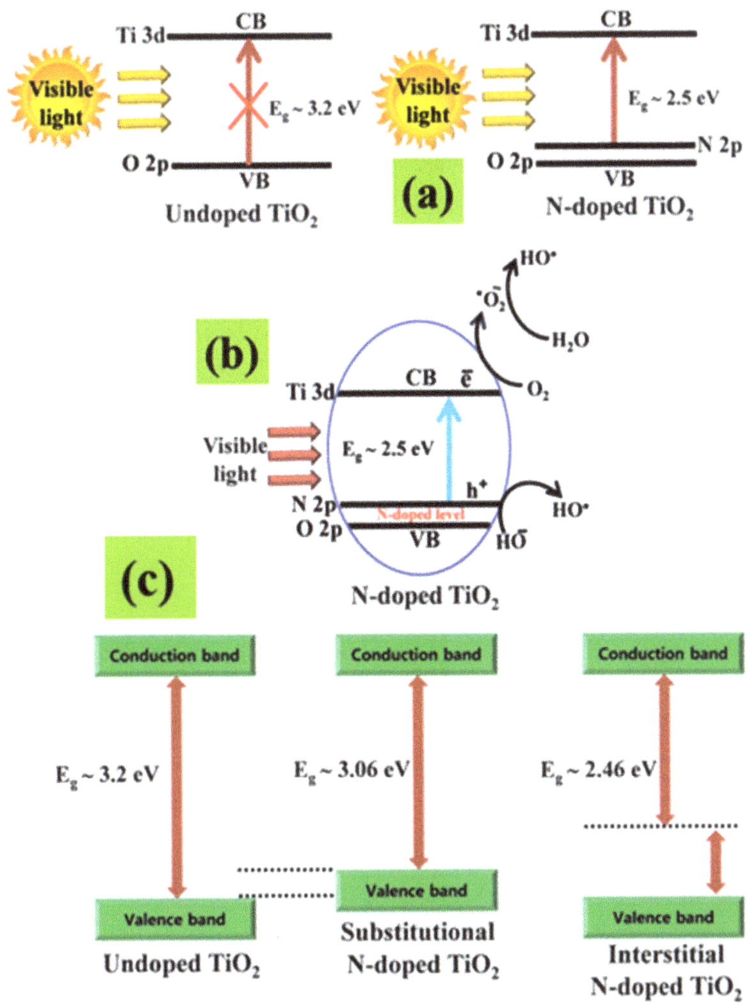

Figure 2.6. (a) Energy level diagrams for undoped and N-doped TiO_2 and (b) possible reaction mechanism for the photocatalytic degradation of organic pollutants by N-doped TiO_2 under visible light irradiation. (c) The valence and conduction bands of undoped and N-doped TiO_2. (Adapted with permission from [115]. Copyright the Royal Society of Chemistry.)

significantly influenced by the placement of N-doping in the material; nevertheless, there is some debate on whether the position is advantageous. Doping TiO_2 lattices with nitrogen often results in a mixture of substitution and interstitial doping, as reported by various researchers [106]. The preparation procedure for a non-metal-doped TiO_2 photocatalyst may be divided into process treatment and post-treatment. Post-treatment refers to the addition of non-metal components to a material after it has been formed into TiO_2, while process treatment refers to the addition of non-metal elements during the synthesis of TiO_2.

Through the hydrothermal technique,Rahbar *et al* [107] were able to generate S and N co-doped carbon quantum dot (CQD)/TiO_2 composites. Acidic AR88 (azo dyes) degraded at 54% when exposed to visible light, which was faster than the rate at which pure TiO_2 degraded. Additionally, they discovered that due to the peculiarities of electron transport, CQDs allowed for the segregation of charges. The surface functional group also increased the photocatalytic efficiency by enabling a greater capacity for adsorption on the photocatalyst surface. Additionally, to facilitate the photocatalytic process, pollutant molecules have been adsorbed on the surface of the photocatalyst. The sol–gel method was used by Li *et al* to synthesize N-doped TiO_2, and the breakdown of methylene orange (MO) dye showed that the rate of degradation of pure TiO_2 was less than 5% after 180 min of exposure to visible light, whereas the rate of degradation of incorporating N-doped TiO_2 as the photocatalyst was above 95% after 90 min of exposure [108]. Because Ti^{3+} and N impurity worked together to reduce the energy gap of N-doped TiO_2, the N-doping significantly improved TiO_2's capacity to degrade MO when exposed to visible light. In most cases, the introduction of N into the TiO_2 structure results in a new mid-gap energy state formation. This new energy state is known as the N 2p band and is located above the O 2p valence band. In the long run, this reduces the band energy gap of TiO_2 (to a value of 2.5 eV), which moves the optical absorption towards the visible light region [105, 109]. Because of this, it is feasible for the electron move from the valence to conduction band when the material absorbs visible light. This causes N-doped TiO_2 to have visible light activity (figure 2.6(a)) [104, 105, 109]. The suggested schematic diagram (figure 2.6(b)) depicts a lower band gap state that occurs as a result of N-doping as well as the photocatalytic breakdown process under the influence of visible light irradiation. The diagram illustrates how visible light causes an electron to form in the conduction band and a hole to form the valence band. These electrons within the conduction band develop on the surface of the catalyst, and then they are scavenged by oxygen molecules from the environment or water to produce highly oxidative species, such as hydroxyl radicals and superoxide radical anions [110–115]. The production of highly oxidative species occurs as a result of the OH^- ions scavenging the holes that are generated in the valence band. Due to their high reactivity, these highly oxidative species, like hydroxyl radicals ($HO^{\cdot}$), are in charge of mineralizing and degrading organic contaminants.

2.6.2 Heterostructuring

Another method that can be used to improve photocatalytic activity is called heterostructuring. This method works by improving charge separation efficiency.

By mixing the semiconductor with non-metals or metals, heterostructures can be produced. By delocalizing the photoexcited charges into a separate carrier and prohibiting them from recombining, these additional energy states help to separate the charges [116]. For example, electrostatic self-assembly has been used to create heterostructures of ZnO/MXene (Nb_2C and V_2C) by Zhou *et al* [117]. He developed a hierarchical ZnO/MXene heterostructure as a photocatalyst for dye degradation by growing chimeric ZnO nanosheets with high carrier mobility on the surface of the accordion-shaped MXene (Nb_2C and V_2C). The ZnO/Nb_2C and ZnO/V_2C hierarchical structures generated after being exposed to UV radiation for 120 min showed outstanding photoactivity in the breakdown of MB; the degradation rate reached 62.62% and 99.53%, respectively. The heterostructures involving ZnO and MXene, which may decrease the carrier transfer path, were most likely responsible for this. This may make it easier for photogenerated electrons to go from the CB of ZnO to the CB of MXene, so greatly increasing the efficiency of photodegradation. The degradation rates persisted at 58.6% and 97.3%, correspondingly, after four cycles, but they only remained at 22.0% for pure ZnO, demonstrating the superior photocatalytic stability of the hierarchical structures of ZnO/Nb_2C and ZnO/V_2C. This results from the ZnO/MXene hierarchical structure developed to offer homogeneous active sites on the surface and a porous structure, inhibiting aggregation.

In a study that was completed just recently Hossain *et al* [118], using a technique known as ultrasound-assisted wet-impregnation, were able to successfully create a CuO/CdS nanocomposite photocatalyst. In this study a p-type CuO with an n-type CdS heterojunction was created to modify the photocatalytic property of the p-type CuO. The development of a nanocomposite was discovered by detailed structural, optical, and morphological analysis. This nanocomposite showed a substantial improvement in the charge carrier separation (between holes and electrons). Under the irradiation of visible light, the photodegradation effectiveness of CuO and CuO/CdS were examined for a variety of dyes such as methyl orange (MO), rhodamine B (RhB), methyl blue (MB), and methylene blue (MLB). The dye degradation efficiency was obtained as follows: 80%, 93%, 83%, and 75%, respectively. Under visible light, the quantum yield for RhB degradation was calculated to be 6.5×10^{-5}. Tests to determine the photocatalyst's reusability found that the CuO/CdS mixture may be recycled up to four times. Utilizing a wide variety of scavengers allowed for the possible pathways for the photodegradation over CuO/CdS nanocomposite to be uncovered and investigated. As a result of this research, it has been demonstrated beyond a reasonable doubt that the margins of the conduction bands in CuO and CdS play a significant role in the production of O_2^-. In the bulk solution, the generated O_2^- was responsible for the degradation of the dye molecules, while the VB location of CuO served as the oxidation site of water.

Under UV irradiation, Wan *et al* investigated the photodegradation of acetone utilizing ZnO–Zn_2TiO_4 with a hexagonal–spinal structure, respectively. The findings indicate that 95% is the highest possible efficiency for degradation[119]. Pinto *et al* studies the *in situ* synthesis of ZnO/Zn_2TiO_4 nanocomposite and evaluate the degradation performances of the nanocomposite using Reactive Black 5 dye [120].

After 60 min of exposure to visible light, the pollutant underwent complete photodegradation and was rendered harmless. They demonstrated that the photo-activity of pure ZnO may be enhanced with the construction of a heterostructure composed of ZnO and Zn_2TiO_4. The researchers Lim *et al* suggested a method for the fine fabrication of multicore–shell ZnO@Zn_2TiO_4 spheres with a thin covering of carbon. Under the illumination of sunlight, the photocatalytic performance of this hybrid catalyst was examined using an organic dye. After sixty minutes, the deterioration reached around 93% [121]. In the most recent study, ZnO–Zn_2TiO_4 (ZTM) material was produced using an innovative synthesis approach based on an ultrasound-assisted polyol-mediated process [122]. The materials were employed as photocatalysts for the photodegradation of clofibric acid as a typical pharmaceutical contaminant. The performance of the photocatalytic reaction was examined under a variety of calcination temperatures, initial pH, catalyst dose, and starting concentration of clofibric acid solution combinations. According to the findings, the hexagonal–tetragonal phases of ZnO–Zn_2TiO_4 that were calcined at 600 °C (ZTM-600) and with an average crystallite size of 97.8 had the highest level of degrading efficiency (99%). The FT-IR analysis showed the major bands that are characteristic of ZnO and Zn_2TiO_4, and the UV–visible DRS confirmed that the UV–visible regions have a greater capability for absorption. When it comes to the process of clofibric acid photodegradation, the photogenerated electrons are the strongest reactive species involved.

References

[1] Nagajyothi P C, Prabhakar Vattikuti S V, Devarayapalli K C, Yoo K, Shim J and Sreekanth T V M 2020 Green synthesis: photocatalytic degradation of textile dyes using metal and metal oxide nanoparticles—latest trends and advancements *Crit. Rev. Environ. Sci. Technol.* **50** 2617–723

[2] Cao Z, Zhang T, Ren P, Cao D, Lin Y, Wang L, Zhang B and Xiang X 2020 Doping of chlorine from a neoprene adhesive enhances degradation efficiency of dyes by structured TiO_2-coated photocatalytic fabrics *Catalysts* **10** 69

[3] Zheng Y, Cao L, Xing G, Bai Z, Huang J and Zhang Z 2019 Microscale flower-like magnesium oxide for highly efficient photocatalytic degradation of organic dyes in aqueous solution *RSC Adv.* **9** 7338–48

[4] Vallejo W, Cantillo A and Díaz-Uribe C 2020 Methylene blue photodegradation under visible irradiation on Ag-doped ZnO thin films *Int. J. Photoenergy* **2020** 1627498

[5] Melinte V, Stroea L and Chibac-Scutaru A L 2019 Polymer nanocomposites for photocatalytic applications *Catalysts* **9** 986

[6] Ola O and Maroto-Valer M M 2015 Review of material design and reactor engineering on TiO_2 photocatalysis for CO_2 reduction *J. Photochem. Photobiol.* C **24** 16–42

[7] M'Arimi M M, Mecha C A, Kiprop A K and Ramkat R 2020 Recent trends in applications of advanced oxidation processes (AOPs) in bioenergy production: review *Renew. Sustain. Energy Rev.* **121** 109669

[8] Arques A, Amat A M, Santos-Juanes L, Vercher R F, Marín M L and Miranda M A 2007 2,4,6-triphenylthiapyrylium cation as homogeneous solar photocatalyst *Catal. Today* **129** 37–42

[9] Brillas E, Sirés I and Oturan M A 2009 Electro-Fenton process and related electrochemical technologies based on Fenton's reaction chemistry *Chem. Rev.* **109** 6570–631

[10] Clarizia L, Russo D, Di Somma I, Marotta R and Andreozzi R 2017 Homogeneous photo-Fenton processes at near neutral pH: a review *Appl. Catal.* B **209** 358–71

[11] Nerud F, Baldrian P, Gabriel J and Ogbeifun D 2001 Decolorization of synthetic dyes by the Fenton reagent and the Cu/pyridine/H_2O_2 system *Chemosphere* **44** 957–61

[12] Xing M, Xu W, Dong C, Bai Y, Zeng J, Zhou Y, Zhang J and Yin Y 2018 Metal sulfides as excellent co-catalysts for H_2O_2 decomposition in advanced oxidation processes *Chem* **4** 1359–72

[13] Pignatello J J, Oliveros E and MacKay A 2006 Advanced oxidation processes for organic contaminant destruction based on the Fenton reaction and related chemistry *Crit. Rev. Environ. Sci. Technol.* **36** 1–84

[14] Qian H, Hou Q, Yu G, Nie Y, Bai C, Bai X and Ju M 2020 Enhanced removal of dye from wastewater by Fenton process activated by core–shell $NiCo_2O_4$@FePc catalyst *J. Clean. Prod.* **273** 123028

[15] Machulek A, Moraes J E F, Vautier-Giongo C, Silverio C A, Friedrich L C, Nascimento C A O, Gonzalez M C and Quina F H 2007 Abatement of the inhibitory effect of chloride anions on the photo-Fenton process *Environ. Sci. Technol.* **41** 8459–63

[16] Baba Y, Yatagai T, Harada T and Kawase Y 2015 Hydroxyl radical generation in the photo-Fenton process: effects of carboxylic acids on iron redox cycling *Chem. Eng. J.* **277** 229–41

[17] Kasiri M B, Aleboyeh H and Aleboyeh A 2008 Degradation of Acid Blue 74 using Fe-ZSM5 zeolite as a heterogeneous photo-Fenton catalyst *Appl. Catal.* B **84** 9–15

[18] Kušić H, Lončarić Božić A, Koprivanac N and Papić S 2007 Fenton type processes for minimization of organic content in coloured wastewaters. Part II: combination with zeolites *Dyes Pigm.* **74** 388–95

[19] Yang X, Yu H, Guo X, Ding Q, Pullerits T, Wang R, Zhang G, Liang W and Sun M 2017 Plasmon-exciton coupling of monolayer MoS_2–Ag nanoparticles hybrids for surface catalytic reaction *Mater. Today Energy* **5** 72–8

[20] Pizarro P, Guillard C, Perol N and Herrmann J M 2005 Photocatalytic degradation of imazapyr in water: comparison of activities of different supported and unsupported TiO_2-based catalysts *Catal. Today* **101** 211–8

[21] Challagulla S, Nagarjuna R, Ganesan R and Roy S 2017 TiO_2 synthesized by various routes and its role on environmental remediation and alternate energy production *Nano-Struct. Nano-Objects* **12** 147–56

[22] Rauf M A and Ashraf S S 2009 Fundamental principles and application of heterogeneous photocatalytic degradation of dyes in solution *Chem. Eng. J.* **151** 10–8

[23] Machado A E H, de Miranda J A, de Freitas R F, Duarte E T F M, Ferreira L F, Albuquerque Y D T, Ruggiero R, Sattler C and de Oliveira L 2003 Destruction of the organic matter present in effluent from a cellulose and paper industry using photocatalysis *J. Photochem. Photobiol.* A **155** 231–41

[24] Khan I, Saeed K, Ali N, Khan I, Zhang B and Sadiq M 2020 Heterogeneous photo-degradation of industrial dyes: an insight to different mechanisms and rate affecting parameters *J. Environ. Chem. Eng.* **8** 104364

[25] Gupta N K, Ghaffari Y, Kim S, Bae J, Kim K S and Saifuddin M 2020 Photocatalytic degradation of organic pollutants over MFe_2O_4 (M = Co, Ni, Cu, Zn) nanoparticles at neutral pH *Sci. Rep.* **10** 1–11

[26] Ma X *et al* 2021 A novel noble-metal-free Mo_2C–In_2S_3 heterojunction photocatalyst with efficient charge separation for enhanced photocatalytic H_2 evolution under visible light *J. Colloid Interface Sci.* **582** 488–95
[27] Wu Z, Chen X, Liu X, Yang X and Yang Y 2019 A ternary magnetic recyclable ZnO/Fe_3O_4/g-C_3N_4 composite photocatalyst for efficient photodegradation of monoazo dye *Nanoscale Res. Lett.* **14** 147
[28] Low J, Jiang C, Cheng B, Wageh S, Al-Ghamdi A A and Yu J 2017 A review of direct Z-scheme photocatalysts *Small Methods* **1** 1700080
[29] Dashtian K, Ghaedi M, Shirinzadeh H, Hajati S and Shahbazi S 2018 Achieving enhanced blue-light-driven photocatalysis using nanosword-like VO_2/$CuWO_4$ type II n–n heterojunction *Chem. Eng. J.* **339** 189–203
[30] Marschall R 2014 Semiconductor composites: strategies for enhancing charge carrier separation to improve photocatalytic activity *Adv. Funct. Mater.* **24** 2421–40
[31] Zhang L and Jaroniec M 2018 Toward designing semiconductor–semiconductor heterojunctions for photocatalytic applications *Appl. Surf. Sci.* **430** 2–17
[32] Pan T, Chen D, Xu W, Fang J, Wu S, Liu Z, Wu K and Fang Z 2020 Anionic polyacrylamide-assisted construction of thin 2D–2D WO_3/g-C_3N_4 step-scheme heterojunction for enhanced tetracycline degradation under visible light irradiation *J. Hazard. Mater.* **393** 122366
[33] Jiang Z, Wan W, Li H, Yuan S, Zhao H and Wong P K 2018 A hierarchical Z-scheme α-Fe_2O_3/g-C_3N_4 hybrid for enhanced photocatalytic CO_2 reduction *Adv. Mater.* **30** 1706108
[34] Chao Y *et al* 2019 Ultrathin visible-light-driven Mo incorporating In_2O_3–$ZnIn_2Se_4$ Z-scheme nanosheet photocatalysts *Adv. Mater.* **31** 1807226
[35] Maeda K 2013 Z-scheme water splitting using two different semiconductor photocatalysts *ACS Catal.* **3** 1486–503
[36] Kim D and Yong K 2021 Boron doping induced charge transfer switching of a C_3N_4/ZnO photocatalyst from Z-scheme to type II to enhance photocatalytic hydrogen production *Appl. Catal.* B **282** 119538
[37] Lee Y, Hwang Y and Chung Y C 2015 Achieving type I, II, and III heterojunctions using functionalized mxene *ACS Appl. Mater. Interfaces* **7** 7163–9
[38] Huang D, Chen S, Zeng G, Gong X, Zhou C, Cheng M, Xue W, Yan X and Li J 2019 Artificial Z-scheme photocatalytic system: what have been done and where to go? *Coord. Chem. Rev.* **385** 44–80
[39] Zielinska M, Wojnowska-Baryla I and Cydzik-Kwiatkowska A 2018 *Bisphenol A Removal from Water and Wastewater* (Berlin: Springer)
[40] Kumar A 2017 A review on the factors affecting the photocatalytic degradation of hazardous materials *Mater. Sci. Eng. Int. J.* **1** 106–14
[41] Nezamzadeh-Ejhieh A and Karimi-Shamsabadi M 2013 Decolorization of a binary azo dyes mixture using CuO incorporated nanozeolite-X as a heterogeneous catalyst and solar irradiation *Chem. Eng. J.* **228** 631–41
[42] Mansouri M, Mozafari N, Bayati B and Setareshenas N 2019 Photo-catalytic dye degradation of methyl orange using zirconia–zeolite nanoparticles *Bull. Mater. Sci.* **42** 1–11
[43] Zhu H, Jiang R, Fu Y, Guan Y, Yao J, Xiao L and Zeng G 2012 Effective photocatalytic decolorization of methyl orange utilizing TiO_2/ZnO/chitosan nanocomposite films under simulated solar irradiation *Desalination* **286** 41–8
[44] Kansal S K, Singh M and Sud D 2008 Studies on TiO_2/ZnO photocatalysed degradation of lignin *J. Hazard. Mater.* **153** 412–7

[45] Mahvi A H, Ghanbarian M, Nasseri S and Khairi A 2009 Mineralization and discoloration of textile wastewater by TiO_2 nanoparticles *Desalination* **239** 309–16
[46] Riga A, Soutsas K, Ntampegliotis K, Karayannis V and Papapolymerou G 2007 Effect of system parameters and of inorganic salts on the decolorization and degradation of Procion H-exl dyes. Comparison of H_2O_2/UV, Fenton, UV/Fenton, TiO_2/UV and TiO_2/UV/H_2O_2 processes *Desalination* **211** 72–86
[47] Hanafi M F and Sapawe N 2020 Effect of pH on the photocatalytic degradation of Remazol Brilliant Blue dye using zirconia catalyst *Mater. Today Proc.* **31** 260–2
[48] Sapawe N, Jalil A A, Triwahyono S, Sah R N R A, Jusoh N W C, Hairom N H H and Efendi J 2013 Electrochemical strategy for grown ZnO nanoparticles deposited onto HY zeolite with enhanced photodecolorization of methylene blue: effect of the formation of SiOZn bonds *Appl. Catal.* A **456** 144–58
[49] Neppolian B, Kanel S R, Choi H C, Shankar M V, Arabindoo B and Murugesan V 2003 Photocatalytic degradation of reactive yellow 17 dye in aqueous solution in the presence of TiO_2 with cement binder *Int. J. Photoenergy* **5** 45–9
[50] Nguyen C H, Fu C C and Juang R S 2018 Degradation of methylene blue and methyl orange by palladium-doped TiO_2 photocatalysis for water reuse: efficiency and degradation pathways *J. Clean. Prod.* **202** 413–27
[51] Hassena H 2016 Photocatalytic degradation of methylene blue by using Al_2O_3/Fe_2O_3 nano composite under visible light *Mod. Chem. Appl.* **4** 176
[52] Reza K M, Kurny A and Gulshan F 2017 Parameters affecting the photocatalytic degradation of dyes using TiO_2: a review *Appl. Water Sci.* **7** 1569–78
[53] Daneshvar N, Salari D and Khataee A R 2003 Photocatalytic degradation of azo dye Acid Red 14 in water: investigation of the effect of operational parameters *J. Photochem. Photobiol.* A **157** 111–6
[54] Zada N, Khan I and Saeed K 2017 Synthesis of multiwalled carbon nanotubes supported manganese and cobalt zinc oxides nanoparticles for the photodegradation of malachite green *Sep. Sci. Technol.* **52** 1477–85
[55] Kiran Avasarala B and Tirukkovalluri S R 2016 Magnesium doped titania for photocatalytic degradation of dyes in visible light *J. Environ. Anal. Toxicol.* **6** 358
[56] Senthilraja A, Krishnakumar B, Subash B, Sobral A J F N, Swaminathan M and Shanthi M 2016 Sn loaded Au–ZnO photocatalyst for the degradation of AR 18 dye under UV-A light *J. Ind. Eng. Chem.* **33** 51–8
[57] Alahiane S, Qourzal S, El Ouardi M, Abaamrane A and Assabbane A 2014 Factors influencing the photocatalytic degradation of Reactive Yellow 145 by TiO_2-coated non-woven fibers *Am. J. Anal. Chem.* **5** 445–54
[58] Azzaz A A, Assadi A A, Jellali S, Bouzaza A, Wolbert D, Rtimi S and Bousselmi L 2018 Discoloration of simulated textile effluent in continuous photoreactor using immobilized titanium dioxide: effect of zinc and sodium chloride *J. Photochem. Photobiol.* A **358** 111–20
[59] Isari A A, Payan A, Fattahi M, Jorfi S and Kakavandi B 2018 Photocatalytic degradation of rhodamine B and real textile wastewater using Fe-doped TiO_2 anchored on reduced graphene oxide (Fe–TiO_2/rGO): characterization and feasibility, mechanism and pathway studies *Appl. Surf. Sci.* **462** 549–64
[60] Sudrajat H and Babel S 2017 A novel visible light active N-doped ZnO for photocatalytic degradation of dyes *J. Water Process. Eng.* **16** 309–18

[61] Ye Y, Feng Y, Bruning H, Yntema D and Rijnaarts H H M 2018 Photocatalytic degradation of metoprolol by TiO_2 nanotube arrays and UV–LED: effects of catalyst properties, operational parameters, commonly present water constituents, and photo-induced reactive species *Appl. Catal.* B **220** 171–81

[62] Dalhatou S, Pétrier C, Laminsi S and Baup S 2015 Sonochemical removal of naphthol blue black azo dye: influence of parameters and effect of mineral ions *Int. J. Environ. Sci. Technol.* **12** 35–44

[63] Mehrotra K, Yablonsky G S and Ray A K 2005 Macro kinetic studies for photocatalytic degradation of benzoic acid in immobilized systems *Chemosphere* **60** 1427–36

[64] Hashimoto K, Irie H and Fujishima A 2005 TiO_2 photocatalysis: a historical overview and future prospects *Jpn. J. Appl. Phys.* **44** 8269–85

[65] Barakat N A M, Kanjwal M A, Chronakis I S and Kim H Y 2013 Influence of temperature on the photodegradation process using Ag-doped TiO_2 nanostructures: negative impact with the nanofibers *J. Mol. Catal.* A **366** 333–40

[66] Shams-Ghahfarokhi Z and Nezamzadeh-Ejhieh A 2015 As-synthesized ZSM-5 zeolite as a suitable support for increasing the photoactivity of semiconductors in a typical photo-degradation process *Mater. Sci. Semicond. Process.* **39** 265–75

[67] Priyanka and Srivastava V C 2013 Photocatalytic oxidation of dye bearing wastewater by iron doped zinc oxide *Ind. Eng. Chem. Res.* **52** 17790–9

[68] Stylidi M, Kondarides D I and Verykios X E 2004 Visible light-induced photocatalytic degradation of Acid Orange 7 in aqueous TiO_2 suspensions *Appl. Catal.* B **47** 189–201

[69] Curcó D, Giménez J, Addardak A, Cervera-March S and Esplugas S 2002 Effects of radiation absorption and catalyst concentration on the photocatalytic degradation of pollutants *Catal. Today* **76** 177–88

[70] Collazzo G C, Foletto E L, Jahn S L and Villetti M A 2012 Degradation of Direct Black 38 dye under visible light and sunlight irradiation by N-doped anatase TIO_2 as photocatalyst *J. Environ. Manage.* **98** 107–11

[71] Abdel-Aziz R, Ahmed M A and Abdel-Messih M F 2020 A novel UV and visible light driven photocatalyst $AgIO_4/ZnO$ nanoparticles with highly enhanced photocatalytic performance for removal of rhodamine B and indigo carmine dyes *J. Photochem. Photobiol.* A **389** 112245

[72] Wu D, Xia K, Fang C, Chen X and Ye Y 2020 Rapid removal of azophloxine via catalytic degradation by a novel heterogeneous catalyst under visible light *Catalysts* **10** 138

[73] Jamil A, Bokhari T H, Javed T, Mustafa R, Sajid M, Noreen S, Zuber M, Nazir A, Iqbal M and Jilani M I 2020 Photocatalytic degradation of Disperse dye Violet-26 using TiO_2 and ZnO nanomaterials and process variable optimization *J. Mater. Res. Technol.* **9** 1119–28

[74] Rahmat M, Rehman A, Rahmat S, Bhatti H N, Iqbal M, Khan W S, Bajwa S Z, Rahmat R and Nazir A 2019 Highly efficient removal of crystal violet dye from water by MnO_2 based nanofibrous mesh/photocatalytic process *J. Mater. Res. Technol.* **8** 5149–59

[75] Shah A R, Tahir H and Yasmeen T 2020 Optimization and modeling of photooxidative process for the degradation of Reactive Red223 *J. Chem. Soc. Pak.* **42** 42–56

[76] Tabaï A, Bechiri O and Abbessi M 2017 Degradation of organic dye using a new homogeneous Fenton-like system based on hydrogen peroxide and a recyclable Dawson-type heteropolyanion *Int. J. Ind. Chem.* **8** 83–9

[77] Inderyas A *et al* 2020 Synthesis of immobilized ZnO over polyurethane and photocatalytic activity evaluation for the degradation of azo dye under UV and solar light irradiation *Mater. Res. Express* **7** 25033

[78] Ashraf M *et al* 2020 Hematite and magnetite nanostructures for green and sustainable energy harnessing and environmental pollution control: a review *Chem. Res. Toxicol.* **33** 1292–311
[79] Bokhari T H, Ahmad N, Jilani M I, Saeed M, Usman M, Haq A U, Rehman R, Iqbal M, Nazir A and Javed T 2020 UV/H_2O_2, UV/H_2O_2/SnO_2 and Fe/H_2O_2 based advanced oxidation processes for the degradation of Disperse Violet 63 in aqueous medium *Mater. Res. Express* **7** 015531
[80] Sonkusare V N, Chaudhary R G, Bhusari G S, Mondal A, Potbhare A K, Mishra R K, Juneja H D and Abdala A A 2020 Mesoporous octahedron-shaped tricobalt tetroxide nanoparticles for photocatalytic degradation of toxic dyes *ACS Omega* **5** 7823–35
[81] Zada N, Khan I, Shah T, Gul T, Khan N and Saeed K 2020 Ag–Co oxides nanoparticles supported on carbon nanotubes as an effective catalyst for the photodegradation of Congo red dye in aqueous medium *Inorg. Nano-Metal Chem.* **50** 333–40
[82] Sahoo A and Patra S 2020 A magnetically separable and recyclable g-C_3N_4/Fe_3O_4/porous ruthenium nanocatalyst for the photocatalytic degradation of water-soluble aromatic amines and azo dyes *RSC Adv.* **10** 6043–51
[83] Pant A, Tanwar R, Kaur B and Mandal U K 2018 A magnetically recyclable photocatalyst with commendable dye degradation activity at ambient conditions *Sci. Rep.* **8** 14700
[84] Durgam K, Eppa R, Ramana R M V, Siva K J and Sayanna R 2019 Structural, optical properties and photocatalytic activity of nanocrystalline TiO_2 thin films deposited by sol–gel spin coating *I-Manager's J. Mater. Sci.* **7** 28
[85] Bhardwaj S, Dogra D, Pal B and Singh S 2019 Photodeposition time dependant growth, size and photoactivity of Ag and Cu deposited TiO_2 nanocatalyst under solar irradiation *Sol. Energy* **194** 618–27
[86] Ghanbari S, Givianrad M H and Aberoomand Azar P 2020 Synthesis and characterization of visible light driven N–Fe-codoped TiO_2/SiO_2 for simultaneous photoremoval of Cr (VI) and azo dyes in a novel fixed bed continuous flow photoreactor *Can. J. Chem. Eng.* **98** 705–16
[87] Ahadi S, Moalej N S and Sheibani S 2019 Characteristics and photocatalytic behavior of Fe and Cu doped TiO_2 prepared by combined sol–gel and mechanical alloying *Solid State Sci.* **96** 105975
[88] Zhang J, Fu D, Wang S, Hao R and Xie Y 2019 Photocatalytic removal of chromium(VI) and sulfite using transition metal (Cu, Fe, Zn) doped TiO_2 driven by visible light: feasibility, mechanism and kinetics *J. Ind. Eng. Chem.* **80** 23–32
[89] Majeed Khan M A, Siwach R, Kumar S and Alhazaa A N 2019 Role of Fe doping in tuning photocatalytic and photoelectrochemical properties of TiO_2 for photodegradation of methylene blue *Opt. Laser Technol.* **118** 170–8
[90] Bhatia V and Dhir A 2016 Transition metal doped TiO_2 mediated photocatalytic degradation of anti-inflammatory drug under solar irradiations *J. Environ. Chem. Eng.* **4** 1267–73
[91] Ratova M, West G T, Kelly P J, Xia X and Gao Y 2015 Synergistic effect of doping with nitrogen and molybdenum on the photocatalytic properties of thin titania films *Vacuum* **114** 205–12
[92] Stucchi M, Boffito D C, Pargoletti E, Cerrato G, Bianchi C L and Cappelletti G 2018 Nano-MnO_2 decoration of TiO_2 microparticles to promote gaseous ethanol visible photo-removal *Nanomaterials* **8** 686
[93] Hoffmann M R, Martin S T, Choi W and Bahnemann D W 1995 Environmental applications of semiconductor photocatalysis *Chem. Rev.* **95** 69–96

[94] Anpo M and Takeuchi M 2003 The design and development of highly reactive titanium oxide photocatalysts operating under visible light irradiation *J. Catal.* **216** 505–16

[95] Li R, Li T and Zhou Q 2020 Impact of titanium dioxide (TiO_2) modification on its application to pollution treatment—a review *Catalysts* **10** 804

[96] Crişan M *et al* 2018 Iron doped TiO_2 films and their photoactivity in nitrobenzene removal from water *Appl. Surf. Sci.* **455** 201–15

[97] Gnanasekaran L, Hemamalini R, Saravanan R, Ravichandran K, Gracia F and Gupta V K 2016 Intermediate state created by dopant ions (Mn, Co and Zr) into TiO_2 nanoparticles for degradation of dyes under visible light *J. Mol. Liq.* **223** 652–9

[98] Wang Q, Jin R, Zhang M and Gao S 2017 Solvothermal preparation of Fe-doped TiO_2 nanotube arrays for enhancement in visible light induced photoelectrochemical performance *J. Alloys Compd.* **690** 139–44

[99] Freitas F S, Gonçalves A S, De Morais A, Benedetti J E and Nogueira A F 2012 Graphene-like MoS_2 as a low-cost counter electrode material for dye-sensitized solar cells *J. NanoGe J. Energy Sustain* 11002–3

[100] Akpan U G and Hameed B H 2010 The advancements in sol–gel method of doped-TiO_2 photocatalysts *Appl. Catal.* A **375** 1–11

[101] Vaiano V, Sacco O, Sannino D and Ciambelli P 2016 N-doped ZnO nanoparticles supported on ZnS based blue phosphors in the photocatalytic removal of Eriochrome Black-T dye *Chem. Eng. Trans.* **47** 187–92

[102] Di Valentin C, Pacchioni G, Selloni A, Livraghi S and Giamello E 2005 Characterization of paramagnetic species in N-doped TiO_2 powders by EPR spectroscopy and DFT calculations *J. Phys. Chem.* B **109** 11414–9

[103] Thompson T L and Yates J T 2006 Surface science studies of the photoactivation of TiO_2—new photochemical processes *Chem. Rev.* **106** 4428–53

[104] Asahi R, Morikawa T, Ohwaki T, Aoki K and Taga Y 2001 Visible-light photocatalysis in nitrogen-doped titanium oxides *Science* **293** 269–71

[105] Dunnill C W and Parkin I P 2011 Nitrogen-doped TiO_2 thin films: photocatalytic applications for healthcare environments *Dalton Trans.* **40** 1635–40

[106] Zhao Z and Liu Q 2008 Mechanism of higher photocatalytic activity of anatase TiO_2 doped with nitrogen under visible-light irradiation from density functional theory calculation *J. Phys. D: Appl. Phys.* **41** 10

[107] Rahbar M, Mehrzad M, Behpour M, Mohammadi-Aghdam S and Ashrafi M 2019 S, N co-doped carbon quantum dots/TiO_2 nanocomposite as highly efficient visible light photocatalyst *Nanotechnology* **30** 505702

[108] Li H, Hao Y, Lu H, Liang L, Wang Y, Qiu J, Shi X, Wang Y and Yao J 2015 A systematic study on visible-light N-doped TiO_2 photocatalyst obtained from ethylenediamine by sol–gel method *Appl. Surf. Sci.* **344** 112–8

[109] Asahi R, Morikawa T, Irie H and Ohwaki T 2014 Nitrogen-doped titanium dioxide as visible-light-sensitive photocatalyst: designs, developments, and prospects *Chem. Rev.* **114** 9824–52

[110] Xu H, Ouyang S, Liu L, Reunchan P, Umezawa N and Ye J 2014 Recent advances in TiO_2-based photocatalysis *J. Mater. Chem.* A **2** 12642–61

[111] Kalathil S, Khan M M, Ansari S A, Lee J and Cho M H 2013 Band gap narrowing of titanium dioxide (TiO_2) nanocrystals by electrochemically active biofilms and their visible light activity *Nanoscale* **5** 6323–6

[112] Khan M M, Ansari S A, Pradhan D, Ansari M O, Han D H, Lee J and Cho M H 2014 Band gap engineered TiO_2 nanoparticles for visible light induced photoelectrochemical and photocatalytic studies *J. Mater. Chem.* A **2** 637–44
[113] Ansari S A, Khan M M, Kalathil S, Nisar A, Lee J and Cho M H 2013 Oxygen vacancy induced band gap narrowing of ZnO nanostructures by an electrochemically active biofilm *Nanoscale* **5** 9238–46
[114] Primo A, Corma A and García H 2011 Titania supported gold nanoparticles as photocatalyst *Phys. Chem. Chem. Phys.* **13** 886–910
[115] Ansari S A, Khan M M, Ansari M O and Cho M H 2016 Nitrogen-doped titanium dioxide (N-doped TiO_2) for visible light photocatalysis *New J. Chem.* **40** 3000–9
[116] Dutta S K, Mehetor S K and Pradhan N 2015 Metal semiconductor heterostructures for photocatalytic conversion of light energy *J. Phys. Chem. Lett.* **6** 936–44
[117] Zhou W, Yu B, Zhu J, Li K and Tian S 2022 Hierarchical ZnO/MXene (Nb_2C and V_2C) heterostructure with efficient electron transfer for enhanced photocatalytic activity *Appl. Surf. Sci.* **590** 153095
[118] Hossain S S, Tarek M, Munusamy T D, Rezaul Karim K M, Roopan S M, Sarkar S M, Cheng C K and Khan M M R 2020 Facile synthesis of CuO/CdS heterostructure photocatalyst for the effective degradation of dye under visible light *Environ. Res.* **188** 109803
[119] Wan L, Li X, Qu Z, Shi Y, Li H, Zhao Q and Chen G 2010 Facile synthesis of ZnO/Zn_2TiO_4 core/shell nanowires for photocatalytic oxidation of acetone *J. Hazard. Mater.* **184** 864–8
[120] Pinto R B, Peralta-Zamora P and Wypych F 2018 Fabrication of ZnO–Zn_2TiO_4 nanocomposite from zinc hydroxide nitrate and its photocatalytic efficiency *J. Photochem. Photobiol.* A **353** 46–52
[121] Lim H S, Lee J S, Lee S, Kang Y S, Sun Y K and Do Suh K 2017 Walnut-like ZnO@Zn_2TiO_4 multicore–shell submicron spheres with a thin carbon layer: fine synthesis, facile structural control and solar light photocatalytic application *Acta Mater.* **122** 287–97
[122] Janani F Z, Khiar H, Taoufik N, Elhalil A, Sadiq M, Mansouri S and Barka N 2022 ZnO–Zn_2TiO_4 heterostructure for highly efficient photocatalytic degradation of pharmaceuticals *Environ. Sci. Pollut. Res.* **1** 1–14

IOP Publishing

Photocatalytic Dye Degradation Using Green Polymeric-based Nanostructures

Principles and applications in wastewater treatment

Muhammad Ikram, Muhammad Ahsaan Bari and Junaid Haider

Chapter 3

Chemical compositions of organic dyes

More than 7×10^5 tons of synthetic dye are produced each year across the industry. Because synthetic dyes may be found in a wide variety of products, including leather accessories, furniture, and even clothing, they play a significant role in our everyday lives. One negative consequence of the extensive use of these dyes is that up to 12% of them are lost as waste during the dying process, and around 20% of that lost dye ends up in the environment. Cationic dyes and anionic dyes are the two types that make up the category of dyes. Cationic dyes are those that have a positive charge, whereas anionic dyes have a negative charge. Cationic dyes are far more carcinogenic and mutagenic than their anionic counterparts, making them a more significant health risk. This chapter provides a comprehensive summary of anionic and cationic dyes, as well as their impact on the surrounding environment, and it does so in a very detailed manner. Even at a concentration of 1 ppm, the presence of dyes in the environment is highly hazardous. In light of the foregoing, it is essential to break down these dyes before releasing them into the environment.

3.1 Dyes

Dyes are colored aromatic organic compounds which absorb light and give color to the visible spectrum [1, 2]. More than 100 000 commercial dyes have been recorded around the world, which amounts to roughly $7 \times 10^8 - 1 \times 10^9$ kg of dye produced per year [3]. It is estimated that 10%–15% of the dyes are lost in the effluent during the process of dying the fabric. In 1856 William Henry Perkin produced and named the first synthetic dye, mauveine (an organic aniline dye) [4]. Dye is applied to substrates in order to give them a permanent color that can withstand fading caused by water, light, oxidizing chemicals, perspiration, and microbiological assault [5]. Because of these benefits, a wide variety of dyes are utilized in various sectors, including those that deal with textiles, food, leather, publishing, cosmetology, pharmaceuticals, polymers, concrete, and the paper industry,

doi:10.1088/978-0-7503-5517-9ch3

for a variety of purposes [6]. These companies produce an enormous quantity of wastewater contaminated with carcinogenic and poisonous dyes, which pollutes water to the point that it is unsafe for human consumption [7]. The textile industry is the sector that utilizes textile dyes the most, and it is also the sector that consumes the most dye overall [8]. Textile dyes are very complex compounds that contain a variety of structural groups. Because dyes may be poisonous and can cause cancer, the presence of dyes in water, even in very low amounts, can cause major health concerns [9]. An immediate and pressing requirement is effectively removing wastewater dyes [10]. This is because dyes in wastewater are challenging to break down in nature (biodegradation), and they also have adverse effects not only on humans but also on the environment.

Chromophores, which are responsible for creating the color, and auxochromes, which improve the dye's affinity to the fibers, are two crucial components that are included inside the dye molecules [11]. Acid dyes, basic dyes, direct dyes, reactive dyes, azo dyes, vat dyes, mordant dyes, sulfur dyes, and disperse dyes are the most common types of dyes employed in the textile sector [12]. Azo derivatives are the most popular category of dyes in the industry [13]. Anionic, nonionic, and cationic dyes are the three different types. Basic dyes are called cationic dyes, and anionic dyes include reactive, direct, and acid dyes.

3.2 Cationic dyes

3.2.1 Malachite green

Malachite green (MG) is a type of triphenylmethane that is soluble in water and looks like a green crystalline powder (figure 3.1). MG is produced by a process that takes place in two stages. The first step is to make leuco malachite green, which may be done by condensing benzaldehyde and dimethylamine. In the final step, the colorless leuco compound is oxidized into the equivalent malachite green cation, which produces malachite green dye [14]. In many different commercial applications, MG is frequently utilized as a raw ingredient as a colorant, making it one of the many dyes available [15]. The use of MG in the production of alcohol and medications as well as the coloring of wool, paper, leather, silk, cotton, and jute is very common [16]. Additionally, it is utilized in several aquaculture businesses all over the world as a

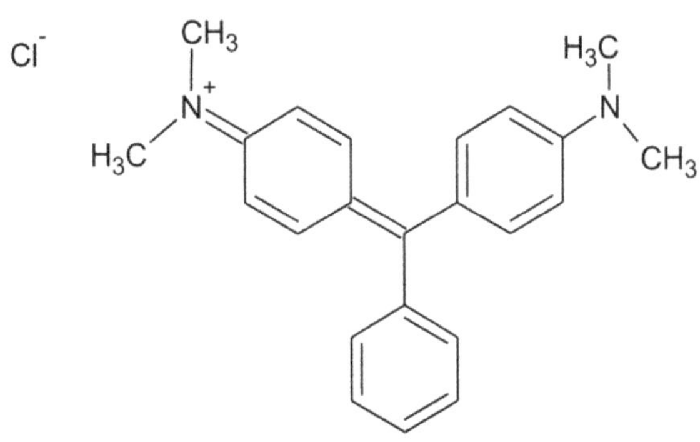

Figure 3.1. The structure of malachite green.

bactericide, fungicide, and parasiticide [17]. It is also used as a food additive, coloring agent, medicinal anthelmintic, and disinfectant [18]. However, MG has been shown in a number of trials to have potentially harmful consequences. It has been found to be a promoter of liver tumors and is particularly dangerous to mammalian cells [15]. MG is a chemical that is known to be carcinogenic, mutagenic, and teratogenic; it is also known to induce toxicity in the respiratory system and harm the reproductive systems of people. In addition, MG has a very high resistance both to light and to reduction caused by oxidizing chemicals [19]. The lungs, skin, eyes, bones, liver, brain, spleen, kidneys, and heart can all suffer damage from this agent [18]. In rats exposed to MG, there was an increased risk of developing a tumor in the ovaries, breasts, and lungs [20]. In response to MG exposure, there was a reduction in hemophilia treatment center (%), Hb (anemia), and red blood cell (RBC) count (dyscrasia), as well as a delay in blood coagulation [21]. There was also an increase in white blood cell (WBC) count (leukocytosis). As a result, it is a management issue of the utmost importance to remove MG from contaminated water before it is dumped into rivers [22].

3.2.2 Methylene blue

Methylene blue (MB) is a powder that is dark green in color, odorless, and solid at ambient temperature. It produces a blue solution when it is mixed with water. At a temperature of 25 °C the molecular diffusivity (D_{mol}) of MB is measured to be 4.7×10^6 ($cm^2\ s^{-1}$). The width of the MB molecule is roughly 9.5 Å, but its length is 13.82 or 14.47 Å. A pKa value of 3.8 has been determined for the MB dye. Water, ethanol, 2-propanol, methanol, ethyl acetate, and acetone are among the solvents that are capable of dissolving it. At a temperature of 25 °C, it is soluble in water at a rate of 43.6 g l^{-1}. MB has a melting point (T_m) that falls between 100 °C and 110 °C [23].

MB is colorless in its reduced form (leucomethylene) and has a distinctive deep blue hue when it is oxidized (figure 3.2). The chromophoric and auxochrome groups

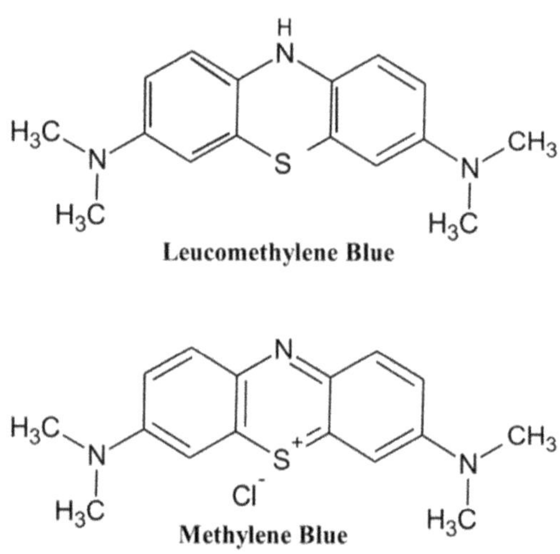

Figure 3.2. The oxidized and reduced forms of MB.

present in MB are responsible for its color. The auxochrome group in MB comprises N-containing groups with lone pair electrons on the benzene ring, while the chromophore group in MB is the N–S conjugated system on the core aromatic heterocycle [24].

MB is an appealing molecule with various features that are helpful for the medical field. It is an effective therapeutic drug used to treat conditions such as anemia, malaria, and Barrett's esophagus. The most common applications for MB have been in human and veterinary medicine, where it has been utilized in various diagnostic and therapeutic treatments. As a photosensitizing agent, it is commonly used to photodynamically inactivate RNA viruses in plasma (such as hepatitis C, HIV, and hepatitis B). Recent research has shown the possibility that MB might help enhance cognition and reduce the symptoms of Alzheimer's disease. Currently, it is being utilized clinically in various drugs that treat illnesses, including ifosfamide-induced encephalopathy, cancer chemotherapy, urinary tract infections, thyroid surgery, plaque psoriasis, and methemoglobinemia. In the research that led to the development of tricyclic antidepressants such as chlorpromazine, MB was the most important molecule. In addition, it has been utilized in diagnosing neuroendocrine tumors, namely insulinomas [23].

The pharmaceutical, textile, printing, medical, paint, food, and paper sectors are only some prospective fields that might benefit from using MB dye [25]. As a dye, it is frequently used in the textile industrial sector and regarded as a popular colorant for clothes. When employed in the textile industry, MB adheres tightly to fabric and clings to the gaps between cotton strands [26]. MB is also used to estimate rock swelling. This information is then used as a quick test to determine the purity of foundry sand in foundries. In analytical chemistry, MB dye is employed not only as a photosensitizer but also as an optical redox indicator, an oxidation–reduction indicator, and for trace identification of anionic surfactants. Additionally, it is evaluated for usage as a candidate material in dye-sensitized capacitors, solar cells, microbial fuel cells, sensors, and other applications [23].

Because of its high toxicity level, MB dye can be hazardous to a person's health when it is present in concentrations over a particular threshold. MB is not only poisonous but also carcinogenic and non-biodegradable, which means it poses a significant risk to human health and negatively impacts the natural environment [27]. MB has been linked to a variety of adverse effects on human health, including but not limited to respiratory discomfort, blindness, gastrointestinal issues, and mental and digestive diseases. Additionally, it causes eye irritation, skin irritation, vomiting, nausea, diarrhea, shock, cyanosis, jaundice, gastritis, tissue necrosis, and methemoglobinemia, which all result in the premature death of cells in tissues [28]. When MB comes into contact with skin, there is a possibility that the skin will become irritated and red. For many reasons, including toxicity and aesthetics, the discharge of MB into the ecosystem raises serious concerns. In addition, it lessens the amount of light that can get through water and is a toxin in the food chains of organisms [29]. Even at extremely low concentrations, the presence of MB in bodies of water produces sub-products with very intense coloration. It reduces oxygen solubility, interferes with aquatic life's ability to photosynthesize, and limits the richness and attractiveness of the natural ecosystem because of its significant molar absorption coefficient (8.4×10^4 l mol^{-1} cm^{-1} at 664 nm), which prevents sunlight penetration [23, 30].

3.2.3 Methyl violet

Methyl violet (MV) is a cationic dye that is a member of the triphenylmethane group (figure 3.3). This dye is frequently utilized in the printing ink, paint industries, and textile for coloring coir, silk, nylon, acrylic, cotton, and wool. Additionally, it has been used to color paper, porcelain, plastic, leather, wax, fur, and soap. In biomedicine MV is the primary component of Gram's biological stain, which is used to classify bacteria. Although it has been shown that most animals are poisoned by this dye, it can occasionally be employed as a medium-class cleanser. The stable MV molecule has a complex chemistry or a long half-life due to the presence of aromatic rings and $-NH_2$ groups, all of which contribute to the compound's ability to survive in the environment. In combustion or oxidation processes, it breaks down into potentially harmful gases, including nitrogen oxide, carbon monoxide, and carbon dioxide. The dye is poisonous, mutagenic, and carcinogenic in its natural state. It can enter the human body through ingestion or inhalation, irritating the eyes and skin, causing vomiting, jaundice cyanosis, shock, and an accelerated heartbeat. MV interacts quickly with the negatively charged cell surface membrane, which allows it to permeate the cytoplasm and deposit itself there. Even worse, it could result in irreversible blindness, the collapse of the respiratory and kidney systems, as well as cancer. Even at concentrations as low as 1 mg l^{-1} the presence of MV in aqueous environments can impair the photosynthesis of aquatic plants since it is a vivid water-soluble dye [31].

3.2.4 Rhodamine B

Rhodamine B (RhB) is a xanthene-based dye that belongs to the basic group and is very soluble in water [32]. The RhB dye family is among the synthetic dyes that have

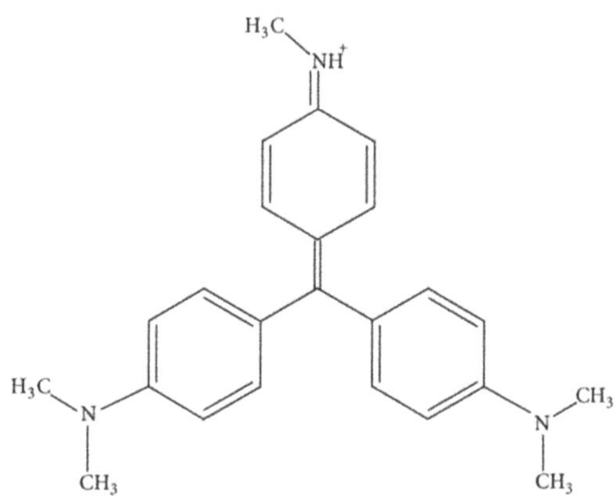

Figure 3.3. Chemical structure of MV 2B dye.

been the most commonly used and oldest, and it has been utilized in various commercial and industrial settings. It is a list I inert ingredient and occurs in three monomeric molecular forms, including lactone, cation, and zwitterion; however, the lactone form does not contribute to the color. Among the basic dyes in the xanthene class, RhB is one of the most eco-friendly options. RhB is a cationic dye that is utilized extensively in the printing business, in the staining of biological studies, in medication (for animals), and in the textile sector. It is known that RhB can cause hazardous and carcinogenic effects, as well as respiratory disorders and renal failure. Because it includes a cationic dye, RhB can irritate human eyes, which can be uncomfortable. RhB endangers aquatic life by reducing the amount of light that can reach through the water's surface, which in turn slows down the process of photosynthesis. Cyprinodon variegates are susceptible to the toxicity of RhB at a concentration of 83.9 mg l^{-1}, which is known as the fatal concentration 50 (LC50) [33]. RhB was first used to impart color to textiles; today, it is also used as a fluorescent marker for microscopic structural tests, water tracing agents, laser dyes, and photosensitizers due to its unique optical qualities [34]. The toxic consequences of RhB on humans and other organisms have been documented; these effects include the development of hypersensitive dermatitis, skin problems, and sickness [35]. Because of RhB's great stability and non-biodegradable nature, it has wide utility in the textile sector as a colorant for textiles [36]. In addition, RhB may be found in products such as paints, ballpoint pens, dye lasers, leather, stamp pad inks, carbon sheets, crackers, and fireworks [37]. Carcinogenic and neurotoxic RhB dye is known to irritate the gastrointestinal tract and respiratory tract as well as cause simulation and developmental toxicity in people and animals. Extensive use of RhB is hazardous when inhaled or ingested, and it also causes damage to both the liver and the thyroid [38].

3.2.5 Crystal violet

As a triphenylmethane dye, crystal violet (CV), also known as (N,N,N1,N1,N11, N11-hexa-methyl-para-rosaniline), has found extensive use as a biological stain in veterinary and human medicine as well as a textile dye in the textile processing industry [39]. The cationic dye CV, commonly known as gentian violet (in its impure form), has one dimethylamine group connected to each phenyl ring. German scientists Kern and Caro first developed CV, which needed the reaction of dimethylaniline with phosgene to develop the intermediate 4,4′-bis(dimethylamine) benzophenone. Then, additional dimethylaniline was added, and the reaction was carried out in the presence of hydrochloric acid and phosphorus oxychloride [40]. Formaldehyde and dimethylaniline may also be condensed together to produce a leuco dye by the condensation process. Next, a suitable oxidizing agent is used to convert this colorless product into a colorful cationic form [41]. This process takes place after the product has been heated. CV is a coloring agent utilized in various industrial treatments, making it one of the most popular. It is often used as a coloring agent for various items, such as leather, paper, detergent, and fertilizer. CV has found great use in textile dyeing, dermatology, and veterinary medicine.

In addition, it may be utilized in producing paints and printing inks. To prevent the spread of fungus, mold, and intestinal parasites, it is added to chicken feed as a feed additive [42]. Given that CV is a protein dye, it may be utilized to improve the quality of bloody fingerprints. It is used as a histology stain in Gram staining, a technique utilized in the medical industry to categorize microorganisms [43]. Additionally, it possesses antibacterial and antifungal capabilities. There is evidence that CV is a recalcitrant dye molecule. This means that it stays in the environment for a long time and can potentially have hazardous effects on aquatic and terrestrial organisms [39]. This dye is both carcinogenic and mutagenic, which poses a significant threat to human health. It is very poisonous to the cells of mammalian organisms, and if it is absorbed in sufficient quantities via the skin, it can irritate both the skin and the digestive system [44]. In humans it has been linked to symptoms including cyanosis, vomiting, eye irritation, jaundice, heart illness, and quadriplegia. In severe situations, it might lead to complications with the respiratory system and the kidneys [45]. Because of these factors, the presence of CV dye in the environment is undesirable, even at a level of 1 ppb, and it needs to be eliminated.

3.2.6 Brilliant green

One of the triphenylmethane cationic dyes, brilliant green (BG) is a crystalline solid that is a bright green in color and odorless [46]. BG is a basic dye used in various industrial and medicinal applications, including in veterinary medicine, and as a dermatological agent, a biological stain, and an addition to chicken feed to inhibit the spread of intestinal parasites, fungi, and mold [47]. In addition, BG is used extensively in the paper and pulp industry as well as the dying industry [47]. Compared to the anionic dyes, the toxicity of the cationic dyes is significantly higher. BG is employed in the production of ink and the staining component for many types of media. It irritates the human gastrointestinal tract, resulting in symptoms such as diarrhea, vomiting, and nausea. It also irritates the human respiratory tract, resulting in shortness of breath and coughing in humans. It is also possible for it to produce dermatitis upon contact with the skin, which manifests as redness and soreness [46]. It is a dye that is extremely poisonous, with the lethal dosage for humans ranging from 50 to 500 mg kg^{-1} of the substance [48]. In addition, research conducted on rats demonstrated that it negatively impacts both the renal and reproductive systems of the animals [49]. In addition, because of the carcinogenic properties of BG dye, its usage is prohibited in a great number of nations [50]. The BG color is harmful if it touches the skin or eyes or causes swallowing problems. If it is breathed in, it will cause damage to the lungs. Carbon dioxide, nitrogen oxide, and sulfur oxides may be produced during its decomposition. Because of this, separating this dye from the effluent is an immediate necessity [51].

3.2.7 Basic fuchsin

Basic fuchsin (BF), also known as complex red phenyl methane dye, comprises primarily three components: magenta II, rosaniline, and pararosaniline. It is characterized as phenyl methane [52]. In addition, BF is a cationic dye that can

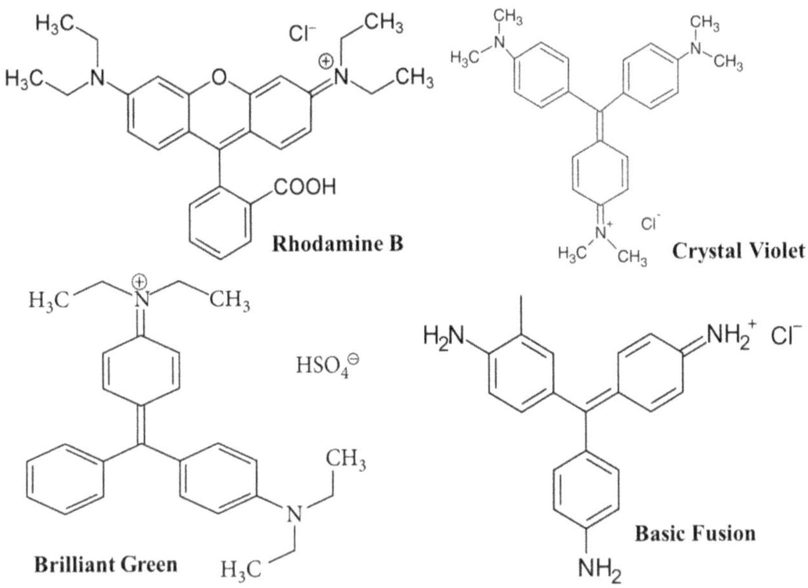

Figure 3.4. Structure of cationic dyes.

also be referred to as rosaniline chloride or basic violet. It is part of the class of triphenylmethane dyes, and its chemical formula is $C_{20}H_{20}ClN_3$. It possesses a remarkable planar conjugated π structure. BF has found applications in various industries, including as biological stains and coloring agents for leather, paper, and cotton [53]. It is used for several purposes, including tracking specific proteins as well as tissue-specific staining of biological and coloring materials such as orlon, nucleus biological stain, muscles, and tissues [54]. In addition, it has been used in research laboratories; for instance, it might be employed as a test molecule in investigations concerning interactions and loadings [55]. In addition, the BF dye can be burned, has anesthetic properties, and inhibits bacteria growth [56]. Because of its toxicity, carcinogenicity, and low biodegradability [57], it has been labeled as a possible carcinogen in several countries. Due to BF's hazardous and carcinogenic effects and low biodegradation, its removal from wastewater becomes crucial to reduce or eliminate it before its discharge into the natural medium [58]. Figure 3.4 demonstrates the structures of the RhB, CV, BG, and BF dyes.

3.3 Anionic dyes

3.3.1 Methyl orange

A common and typical azo-anionic dye is methyl orange (MO), also known as dimethylaminoazobenzenesulfonate. MO was first developed in the 1930s. When dissolved in water, this synthetic organic dye has a greater color ability and appears vibrant orange [59]. This dye is water-soluble and organic. In a titration, the azo dye known as MO is frequently utilized as a pH indicator. This is because it exhibits a visible and distinct color variation when exposed to a range of pH levels. As a pH

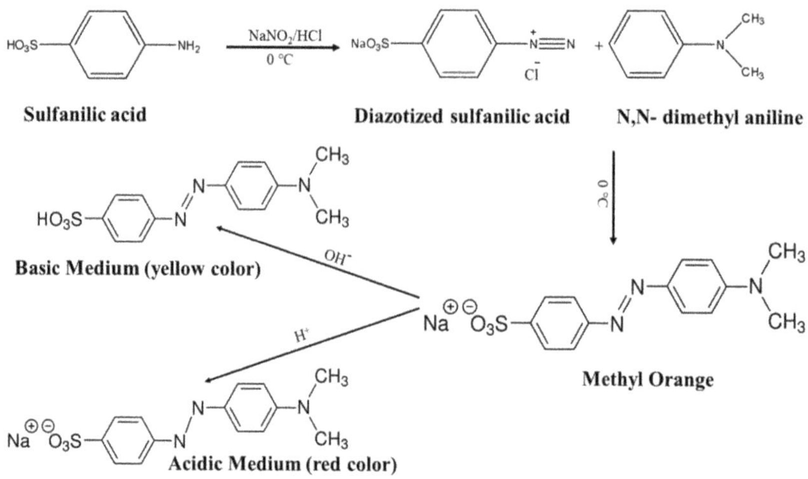

Figure 3.5. Preparation of methyl orange.

indicator, MO exhibits a red color in media with a pH value of less than 3.1, an orange color in media with a pH value between 3.1 and 4.4, and a yellow color in media with a pH value of more than 4.4. One of the steps in the mechanism of producing MO from sulfanilic acid is the diazonium coupling reaction. For this reason, diazotized sulfanilic acid is produced by the reaction of sulfanilic acid with sodium nitrate at 0 °C in the presence of hydrochloric acid. Following this, the diazotized sulfanilic acid is reacted at 0 °C with N,N-dimethylaniline to produce MO (figure 3.5). In an environment with a high acid concentration, molecular oxygen acquires a proton from the acid at its core nitrogen atom, which then undergoes reduction. It brings about a transition from orange to red in hue (figure 3.5). When exposed to a basic medium, MO will acquire an OH^- ion, which will result in the formation of a yellow hue (figure 3.5) [60]. MO is utilized in numerous sectors, such as the textile industry, pharmaceutical industry, printing industry, paper production industry, food industry, and research laboratories. MO causes allergic reactions, sensitization, and dermatitis when breathed in. MO has mutagenic qualities, resulting in short-term or long-term alterations in our DNA that can cause malignancies. As a result, in some circumstances it can also be carcinogenic [60]. It unintentionally gets into the body through eating, and intestinal microbes turn it into aromatic amines, which can cause intestinal cancer [61].

3.3.2 Congo red

Congo red (CR) is an azo dye, an anionic pollutant with a complicated aromatic structure [62]. It is also known as 3,30-([1,10-biphenyl]4.40-diyl)bis(4-amino-1-aminonaphthalene sulfonic) and has the chemical formula $C_{32}H_{22}N_6Na_2O_6S_2$ with a molecular weight of 696.7 g mol^{-1} [63]. Because of the features of the π-conjugated azo link and its resonance, CR dye is particularly robust to light and harsh conditions [64]. As a result, it can withstand exposure to air and light without degrading. The most common reasons for the use of CR are the fact that it is simple to use, inexpensive, and has excellent adhesion. In laboratories CR dye is used to diagnose amyloidosis, as a histology dye and pH indicator, and to evaluate the amount of HCl in the gastrointestinal contents. Because of its strong affinity to the cellulose fibers used in the textile, papermaking, plastics, cosmetics, and pharmaceutical sectors [65], this dye is also utilized in various other commercial applications. Eutrophication, a significant environmental risk, and disorders such as somnolence, which put human life in danger, are also potential effects of CR [63]. CR is a member of the benzene compounds and incorporates aromatic amines into its structure; as a result, it is likely to be transformed into carcinogenic materials when exposed to the environment. In addition, the benzidine formed as a byproduct of the metabolism of CR is known to have hazardous effects on people's health, including the irritation of the skin, the eyes, and the gastrointestinal tract. In addition, benzidine contributes to the coagulation of blood, somnolence, and breathing difficulties [65]. It is extremely soluble in water, does not degrade when exposed to typical degradation methods, and does not degrade when exposed to biological degradation [62]. Based on the pH, different CR molecule configurations may be available in an aqueous solution, making the removal process more challenging [66]. In particular, it has a high sensitivity to acids and when exposed to them its color changes from red to blue (below pH 5). The variations in hue could be caused by the protonation of its amino groups or resonance between charged canonical structures [67]. As a result, it is necessary to remove CR from the sewage of industries and laboratories before it is released into the aquatic environment.

3.3.3 Alizarin Red S

Anthraquinone dyes, such as Alizarin Red S (ARS), are anionic and have a long shelf life. Sulfonation is the chemical reaction used to produce ARS from alizarin, a natural dye whose origin is the madder plant (*Rubia tinctorum*, L. *rubiaceae*) [68]. In therapeutic settings, this dye is used to stain synovial fluid in order to examine it for the presence of basic calcium phosphate crystals. Geologists use it as a staining agent to distinguish calcium carbonate minerals as aragonite and calcite [69]. Other popular names for this dye are diamond red-W, Alizarin Red, Alizarin Carmine, and Mordant Red-3 water soluble [69]. It is a powder that looks orange and has no smell. ARS is a well-known complexing agent for conducting analytical research on metal ions. ARS is capable of forming two distinct forms of complexes with metal ions, which are referred to as monometallic (colored red) and bimetallic (colored blue) [70]. In the textile industry, it is most frequently put to use as a staining agent [68]. It is said to be one of the most difficult to eradicate and long-lasting pollutants. This results from its intricate structures consisting of aromatic rings, which provide a high level of optical and physicochemical stability [71]. Those who suffer from skin allergies are at a greater risk of experiencing its harmful consequences. Because of its acute toxicity it can irritate a person's eyes, skin, mucous membranes, lungs, and digestive tract. Dermatitis develops in patients with chronic diseases. This dye is also utilized to stain biological materials such as the calcified bones of vertebrate groups and tiny embryos of invertebrate species. Because this man-made dye is a powerful oxidizing agent, it must be kept in a cool, dry place away from any heat sources. ARS is a chemically stable dye that does not biodegrade; as a result, it cannot be mineralized by using traditional treatment techniques [68].

3.3.4 Eosin Y

Eosin Y (EY), a man-made heterocyclic chemical, is a triclinic crystal-based red fluorescent dye. All forms of eosins are bromine derivatives of fluorescein [72]. Numerous industries use it, including for biological research, printing inks, leather, coloring, fluorescent pigments, and diagnostic purposes [73]. They find use in the coloring of cosmetics and fuels, as well as in the dyeing and textile industries. It has been demonstrated that EY, a stable dye, can act as a photocatalyst for the degradation of several aromatics, including arenediazonium salts. The dye exhibits redox characteristics with a redox potential of 1.1 V (versus saturated calomel electrode) for the pair EY^{+}/EY^{*} [74]. Because of its great capacity for robust absorption by red blood cells and its red hue, it is suitable for gram staining, distinguishing various species of bacteria [72]. Effluent that contains EY is a significant environmental risk because it is poisonous to people and animals, is also known to be mutagenic, and appears as a dull yellow tint [73]. Because of the dye's potentially harmful effects, the United States government has made its use in edible products illegal. The EY exhibits toxicological concerns such as severe eye and skin irritation with discomfort and redness, and repeated consumption causes negative effects primarily on the vital organs such as the kidney, liver, etc, as well as significant corneal injury by damaging the ganglion cells. Additionally, it alters the DNA of gastrointestinal organs in living organisms, which causes a variety of disorders in humans. After being broken down by heat, an oxidizing agent, and light, the dyes' metabolites are also extremely poisonous and have cancer-causing properties [75].

3.3.5 Orange II

The anionic dye known as orange II (OII), also known as acid orange 7, is a member of the azo group [76]. OII is classified as an azo dye, accounting for more than 15% of the world's output of dyes used in the textile manufacturing business [77]. OII is a well-known water-soluble dye used for coloring a wide range of materials, including aluminum, nylon, silk, detergents, wool, and cosmetics, among other things. As with the vast majority of other azo dyes, it is often disposed of in industrial effluent and poses a significant risk to human health. Its intake can lead to irritation of the skin, eyes, upper respiratory tract, and mucous membranes, nausea, severe headaches, loss of bone marrow leading to anemia and water-borne diseases such as dermatitis. Consuming it poses an additional risk of death because it is a carcinogen and can potentially cause tumors. It is now generally known that the azo group's electron-withdrawing properties are the main source of its long-term toxicity. When an azo group lacks electrons, it transforms into carcinogenic amino compounds. This theory has been around for quite some time and has received much support. It was established that reducing OII results in the production of 1-amino-2-naphthol, which is known to cause bladder cancers. OII can quickly undergo enzymatic breakdown in addition to reduction and cleavage, which, with exposure, can induce methemoglobinemia. Aromatic amines are produced as a byproduct of this process. Additionally, the intermediate amines produced in this manner are predisposed to oxidize hemoglobin's heme iron from Fe(II) to Fe(III) and inhibit oxygen binding, which causes the onset of various usual symptoms such as cyanosis of the lips and nose, dizziness, and weakness. OII is considered genotoxic when consumed by humans, however, it has mutagenic activity in the presence of specific contaminants, such as aromatic amines [78]. It is not affected by the light's deterioration, the action of oxygen, or ordinary acids or bases. OII does not go through the process of biological degradation in wastewater treatment plants [77]. The stability of OII is advantageous for manufacturing textiles, but it makes removal difficult to manage.

3.3.6 Rose bengal

Fluorescein is a type of dye that includes halogens, and rose bengal (RB), also known as acid red 94, is its popular name. It is a water-soluble dye with a pink hue [79]. The cosmetics, plastics, printing, and textile sectors are some of the most common end users of RB dyes [80]. Staining necrotic tissue in the eyes and devitalized corneal cells in conjunctivitis are two of its many diagnostic applications in medicine [79]. Other diagnostic applications include testing liver function. Because RB is capable of transmitting long-term radicals from the triplet excited state, it is susceptible to intersystem transitions. Because of its high responsiveness to the immediate microenvironment, it is an excellent choice for photophysical research [81]. RB, as a sensitizer for the photogeneration of singlet oxygen ($^1O_2$), has seen widespread use in photochemical processes and photodynamic treatment [82]. Because of its many varied features, RB has come to play a unique role in various industries, most notably in the medical industry, where it is extensively utilized in the ophthalmological diagnosis of the disease commonly referred to as 'dry eye'. Recent research has shown that the dye can be used in treatment methods that include cancer cells [83]. A low-concentration dose of RB is also utilized in the treatment of psoriasis and eczema, respectively. In addition to its various therapeutic uses, RB has significant deleterious effects on the human corneal epithelium [79]. When the dye contacts skin, it can cause itching, irritation, reddening, scaling, and even blistering in rare cases. Another set of symptoms includes inflammation, redness, watery eyes, and itching in the eye area. People who inhale it experience respiratory pain due to the destruction of their mucous membranes [84]. It inhibits the formation of bacterial colonies and slows the spread of mold spores throughout the environment. Additionally, it is utilized to assess the kind of nucleotide binding sites in some enzymes, photoinactivation, and identify amino acid residues at the binding sites of some enzymes [81].

3.3.7 Eriochrome Black T

Eriochrome Black T (EBT) is an anionic azo dye most commonly used in the dying of textiles, including nylon fibers, wool, and silk [85]. In the textile industry and research facilities, this dye has been utilized as an indicator in titration and, more recently, complexation [86]. It is also utilized as an indicator in complexometric titration, which is used to determine a substance's level of hardness. EBT is a part of the azo dye group, which accounts for seventy percent of all dyes manufactured commercially and utilized in the industry [87]. Additionally, pure EBT is utilized as an indicator in complexometric titrations for detecting Zn^{2+}, Mg^{2+}, Ca^{2+}, and ions, in addition to being utilized for biological staining [88]. Because of the chemical stability of EBT, it has been categorized as a hazardous dye. It has been shown that EBT is harmful to aquatic creatures and may have long-term negative impacts on aquatic ecosystems. The fact that the EBT structure contains sulfonate groups is a potential source of danger for both humans and the aquatic environment [85]. The dye is dangerous, and its breakdown metabolites, such as naphthoquinone, are much more carcinogenic [88]. Despite this, EBT is notoriously difficult to extract from aqueous solutions due to its molecular structure. When EBT is used in the treatment of polluted wastewater with particular chemical reagents, it forms more complicated chemical structures, which in turn gives it a stronger resistance to photodegradation [86]. Because of its great water solubility and chemical stability, extracting EBT from water is a task that is notoriously challenging [89].

3.3.8 Brilliant Blue-R

Remazol Brilliant Blue-R (RBBR) is a recalcitrant organic pollutant that belongs to the brilliant blue family and is a triphenylmethane dye. It is the most significant dye utilized in the leather and textiles industries, as well as protein quantification, electrophoresis, biochemical analysis, and other applications. As a starting material, it plays an important role in producing polymeric dyes [90]. Because RBBR dye is highly toxic, resistant to biodegradation, highly visible due to its bright color in effluent 'even at a deficient concentration of less than 1 ppm', harmful to aquatic life, and has cancer- and mutagenic-causing properties, discharging it into water bodies poses a severe environmental risk [91]. This risk is compounded by the fact that the dye is highly visible because of its bright color in the effluent. Because of its high level of solubility in aqueous solutions, RBBR, a chemical comprising anthracene derivatives, finds widespread application in the textile sector [92]. RBBR and the products of its decomposition have the potential to cause a variety of adverse health consequences, including irritation of the respiratory and digestive tracts, hyper-activity in children, various forms of allergic reactions, irritation of the eye and skin, asthma, migraines, and so on [91]. The structures of the above-described dyes are provided in figure 3.6.

3.3.9 Indigo carmine

Indigo carmine (IC) is one of the most widely used dyestuffs, and it is naturally derived from the leaves of the *Indigofera tinctoria* plant [93]. It is widely used as a coloring ingredient in various fields [94]. Adolf von Baeyer, a German chemist, discovered the indigo dye's chemical composition in the 1880s. Following the development of that discovery, the first synthetic (man-made) indigo dye suitable for

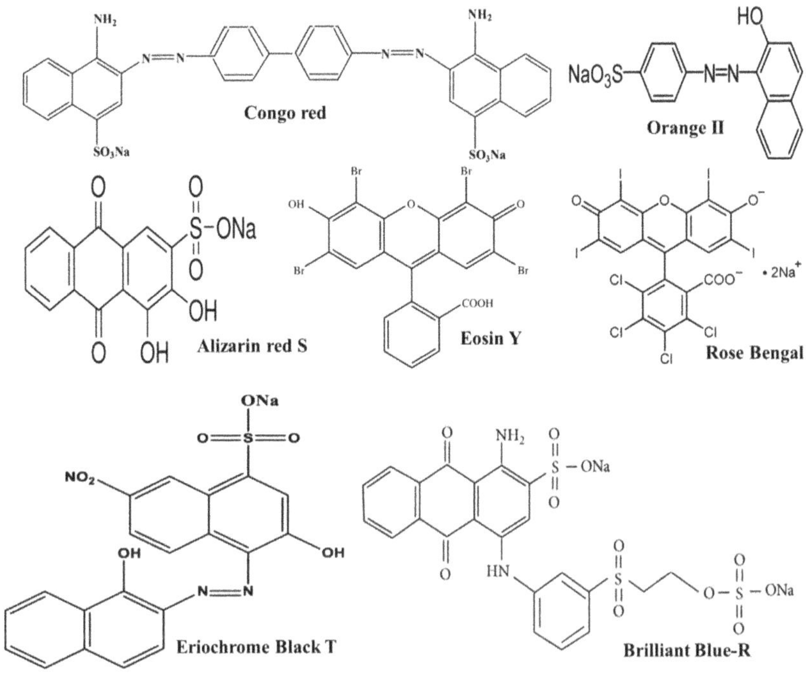

Figure 3.6. Structure of anionic dyes.

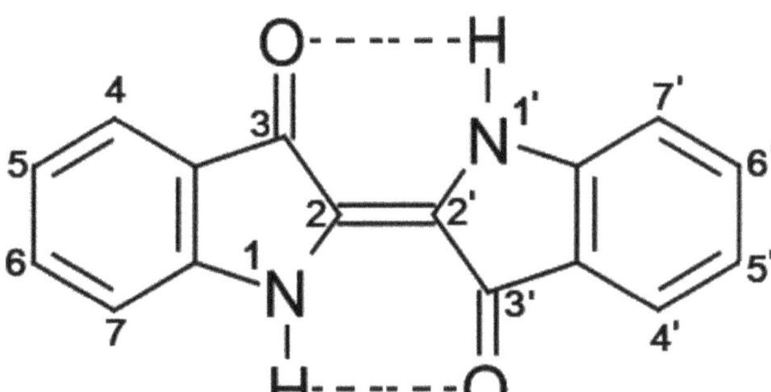

Figure 3.7. General structure of indigo dye. (Reproduced with permission from [96]. Copyright 2020 Elsevier.)

commercial uses was created. Indigo dye is an organic molecule with a characteristic blue hue [95]. Its chemical formula is $C_{16}H_{10}N_2O_2$. Figure 3.7 illustrates the general structure of the indigo dye molecule for clarity.

Indigo dye is almost completely insoluble in water, aqueous acids or bases, and non-polar solvents. However, it only has a trace amount of solubility in dimethylformamide, phthalic anhydride, boiling aniline, nitrobenzene, phenol, and dimethyl sulfoxide [97]. The leuco form of indigo can be produced from insoluble indigo using reducing agents such as sodium dithionite, hydrogen, electrochemical, and hydroxyacetone processes [98]. After that, the leuco form of indigo can dissolve in a sodium hydroxide solution, ultimately forming the disodium salt. In its soluble state, vat indigo binds to the surface of the cellulosic fiber and penetrates it. Modifying the aromatic rings allows the producton of a wide array of indigo derivative dyestuffs. As an illustration, the indigo dye thioindigo ($C_{16}H_8O_2S_2$), which is deep red in color, is produced when the two NH groups of an indigo atom are swapped out for S atoms (figure 3.8(a)). Other common derivatives of indigo are 4,7,4,7-tetrachloro thioindigo (figure 3.8(c)) (by direct chlorination of this indigo), indigo carmine (figure 3.8(b)) (sulfonation products of indigo), and tyrian purple (6,6-di-bromoindigo) (figure 3.8(d)) [99].

The textile sector has a significant need for IC, particularly for the coloring of denim-based products. Indigo dye requires anything from 3 to 12 g to properly color a single pair of blue denim jeans in the textile and denim dyeing industries [100]. Therefore, companies that deal with the dying and washing of textiles are one of the primary sources of unfixed indigo dyes that are discharged into the environment. In addition to its usage in the denim dyeing process for textiles, IC finds widespread use in the fields of analytical chemical technology and biological detection, as well as the pharmaceutical, cosmetic, and food sectors [101]. According to some reports, the yearly production of indigo dyes is somewhere about 40 000 metric tons, which is dependent on the demand from throughout the world [102].

When it comes to the health of humans, releasing IC dyes into the environment is an exceedingly risky endeavor since these colors are extremely toxic when they touch the human body. If indigo dye comes into close contact with the eyes, it can cause irreversible damage to the cornea and the conjunctiva. Indigo is extremely hazardous to human skin and the respiratory tract and causes significant irritation to human skin [103]. Indigo dyes are finding more and more uses in the textile industry due to the growing number of denim fabric mills in Asia, particularly in Pakistan, India, Bangladesh, and China. The widespread use of indigo dye has resulted in substantial contamination across various environmental domains, most notably the water system. The presence of dyestuff in surface water can raise the turbidity of the water, which in

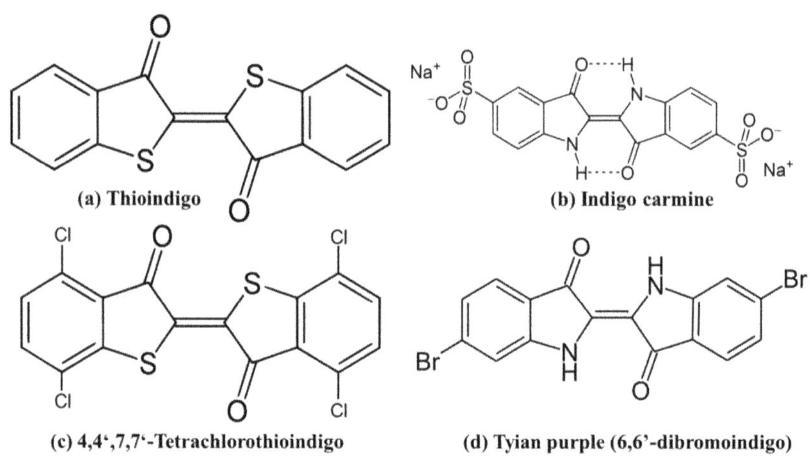

Figure 3.8. Different derivatives of indigo dyes. (Reproduced with permission from [96]. Copyright 2020 Elsevier.)

turn can slow down the photosynthetic process and pose a risk to aquatic ecology [104]. Along with the unfixed dye, a significant amount of alkali and reducing agents are also continually lost as effluents throughout the indigo dyeing process. The effluents that are released into the environment are dangerous to the health of every living thing there. Therefore, in order to safeguard the environment, it is of the utmost importance to separate the indigo color materials from the wastewater.

3.4 Comparison of cationic and anionic dyes

A family of water-soluble dyes known as cationic dyes—often referred to as basic dyes—produce colored cations in solution because their molecules have a delocalized or localized positive charge [105]. They come in various chemical structures founded on substituted aromatic groups [106]. The following are some of the most important chemical categories of cationic dyes: acridine, triarylmethane, oxazine, hemicyanine, cyanine, thiazine, and diazahemicyanine [107]. Because they include aromatic and various functional groups, which, in turn, give the dyes a resistant action, cationic dyes are extremely stable [108]. Cationic dyes, which include all currently available synthetic dyes, are the most brilliant type of water-soluble colors utilized in the textile industry [109]. They are utilized to a large extent in the dyeing of nylon, acrylic, wool, and silk. Hazardous colorants are known to be present in this group of dyes, and they have been connected to a number of harmful health effects, such as skin rashes, allergic dermatitis, cancer, and genetic abnormalities [110]. Cationic dyes have a high tinctorial strength, which may produce a noticeable color even at concentrations as low as 1 ppm [111]. Due to their electrostatic attraction, they quickly and efficiently adsorb onto the negatively charged surface of adsorbents. However, the surface area and functional groups available on the surface of the adsorbent would affect adsorption effectiveness differently [112]. In the context of dye adsorption studies, typical cationic dyes, including malachite green, crystal violet, and methylene blue, have all been used widely as model pollutants. Table 3.1 presents the properties of cationic dyes.

A dye with a negative charge is called an anionic dye. Anionic dyes comprise an extensive range of chemical compounds from several dye classes. Although these compounds range significantly in their structural make-up (such as anthraquinone, azoic, nitro, and triphenylmethane dyes), they all feature water-soluble ionic substituents as part of their chemical composition. Additionally, the group of anionic dyes includes direct dyes, and from a chemical standpoint the class of anionic azo dyes encompasses a substantial part of reactive dyes [113]. Most reactive dyes contain a reactive component that forms covalent bonds when interacting with other materials such as cotton, wool, etc. Due to the hydrolysis of reactive groups in the water phase, reactive dyes have a low degree of fixing, making their release into the environment undesirable [114]. Wool, silk, polypropylene, modified acrylic, and polyamide are some of the fibers that may be dyed with acid dyes. Acid dyes have high water solubility, but because they are organic sulfonic acids, they are toxic to humans [115]. The following is a full and detailed description of anionic dyes along with their associated dangers. Table 3.2 represents the properties of anionic dyes.

Table 3.1. Properties of cationic dyes.

Dye	Molecular formula	Molecular weight (g mol^{-1})	Maximum absorption (nm)	Solubility	Melting point (°C)
Malachite green	$C_{23}H_{25}ClN_2$	364.911	616–620	Very soluble in water, alcohol, methanol, and amyl alcohol.	158–160
Methylene blue	$C_{16}H_{18}ClN_3S$	319.9	664	Soluble in water, chloroform, glacial acetic acid, ethanol, and glycerol; slightly soluble in pyridine.	100–110
Methyl violet	$C_{24}H_{28}N_3Cl$	393.96	587	Soluble in water and ethanol; insoluble in xylene.	137
Rhodamine B	$C_{28}H_{31}ClN_2O_3$	479.02	550	Soluble in water.	210–211
Crystal violet	$C_{25}H_{30}ClN_3$	407.979	590	Soluble in chloroform and water (50 mg ml^{-1} at 27° C); partially soluble in alcohol and glycerol.	205
Brilliant green	$C_{27}H_{34}N_2O_4S$	482.63	623	Soluble in water or alcohol with a green color.	210
Basic fuchsin	$C_{20}H_{20}ClN_3$	337.86	549–552	Soluble in alcohol and acids; slightly soluble in water.	200

Table 3.2. Properties of anionic dyes.

Dye	Molecular formula	Molecular weight (g mol^{-1})	Maximum absorption (nm)	Solubility	Melting point (°C)
Methyl orange	$C_{14}H_{14}N_3NaO_3S$	327.33	464	Slightly soluble in water; easily soluble in hot water; insoluble in ethanol.	300
Congo red	$C_{32}H_{22}N_6Na_2O_6S_2$	696.66	498	Water soluble; better solubility in organic solvents such as ethanol.	360
Alizarin Red S	$C_{14}H_7NaO_7S$	342.253	445	Soluble in water and ethanol.	279–283
Eosin Y	$C_{20}H_6Br_4Na_2O_5$	691.86	517	Soluble in water, alcohol, methanol, ethanol, and ethylene glycol.	295.5
Orange II	$C_{16}H_{11}N_2NaO_4S$	350.32	484	Water soluble.	164
Rose bengal	$C_{20}H_2Cl_4I_4Na_2O_5$	1017.64	549	Water soluble.	>300
Eriochrome Black T	$C_{20}H_{12}N_3NaO_7S$	461.380	566	Water soluble (50 g l^{-1} at 20 °C).	N/A
Indigo carmine	$C_{16}H_8N_2Na_2O_8S_2$	466.36	610	Water soluble (10 g l^{-1} at 25 °C).	390
Brilliant Blue-R	$C_{22}H_{16}N_2Na_2O_{11}S_3$	626.53	595	Water soluble.	N/A

3.5 Environmental concerns of hazardous dyes

There is no evidence to support the hypothesis that the majority of the dyestuffs that are now utilized in the process of dyeing and finishing textiles pose a threat to human health when exposure is at the levels that are typically experienced by factory employees. However, if exposed to them over a prolonged period of time or by accident, there is a chance that they can cause health problems; as a result, dyes and chemicals of any kind need to be handled with extreme caution. Inhaling dye particles can cause various respiratory issues, making this the most prevalent risk associated with reactive dyes. Sometimes they can impact a person's immune system and, in extreme situations, this might mean that the person's body can respond dramatically the next time they inhale the dye. This condition is known as respiratory sensitization, and symptoms include itching, sneezing, and watery eyes, in addition to signs of asthma such as wheezing and coughing. Exposure to chemicals that might function as irritants is likely the cause of most of the health concerns associated with the dyeing and finishing procedures. These substances have the potential to irritate the skin, make the nose itchy or clogged, cause sneezing, and aggravate eye pain. They consist of resins based on formaldehyde, ammonia, acetic acid, various compounds used in shrink-resistant coatings, some optical whiteners, bleach, soda ash, and caustic soda. It has been established that particular reactive, vat, and dispersion dyes can cause skin sensitivity. The textile industry generates a significant quantity of waste in liquid form. The effluents from the textile industry include both organic and inorganic chemicals [116]. A percentage of the dyes applied to the clothes do not become permanently affixed to the fabrics, and there is always a component of these dyes that stays unfixed to the garments and is washed out. It has been discovered that textile effluents include large amounts of these dyes that have not been treated [117]. The amount of water consumed and discharged during the production of different types of textiles can be quite different. It takes around 0.08–0.15 m^3 of water to manufacture 1 kg of cloth, depending on the kind of fabric. After processing between 12 and 20 tons of textiles per day, it is anticipated that between 1000 and 3000 m^3 of water will be released into the environment [118]. These effluents include a high concentration of dyes and chemicals, some of which are not biodegradable and are known to cause cancer; as a result, they represent a significant risk to human health as well as to the environment. These effluents have been treated with various techniques, including flocculation, trickling filters, and electrodialysis, that fall under primary, secondary, and tertiary treatment categories. However, it has been shown that these treatments are not successful in removing all the dyes and chemicals utilized [119]. Not only do the effluents have a high concentration of the dyes used in the industry, but they also contain the chemicals utilized in the various phases of the processing. Trace metals such as chromium, arsenic, copper, and zinc are present in these effluents. These trace metals can potentially cause various health issues, including skin ulceration, hemorrhage, nausea, severe irritation of dermatitis, and skin. It has been discovered that textile effluents contain a variety of different organic and microbiological contaminants [120]. Over the last century, there has been a steady rise in cotton applications [121]. Azo dyes, one of the major categories of synthetic colorants used in the industry, are the most common type of synthetic colorant used to dye cotton fibers. It is difficult to break down azo dyes using the typical treatment methods that are already in use. They are distinguished by the existence of the nitrogen–nitrogen bond (–N=N–) at the core of the molecule, which explains why they have a very low number of electrons [116]. It has been discovered that azo dyes are complex in their natural state, and reductive cleavage testing has revealed that they have carcinogenic properties. These dyes have the potential to change the chemical and physical qualities of the soil, degrade the quality of water bodies, and affect both the flora and wildlife that live in the surrounding environment [122]. It has been discovered that the poisonous nature of dyes promotes the death of soil microorganisms, which in turn affects agricultural production [123].

The presence of azo dyes in water in concentrations of less than one part per million (ppm) is easily detectable [124]. Aesthetic value, transparency, and water–gas solubility are all affected by this. A drop in photosynthetic activity, which leads to a lack of oxygen and deregulates the biological cycles of aquatic biota, is caused when the amount of light that can pass through water is reduced [125]. Many azo dyes are also very toxic to the environment and mutagens, which means that organisms exposed to them can experience consequences ranging from acute to chronic, depending on the length of the exposure and the azo dye concentration.

The primary azo dyes of 1,4-diaminobenzene can induce contact dermatitis, skin irritation, chemosis, rhabdomyolysis, exophthalmos, lacrimation, acute tubular necrosis supervenes, vomiting gastritis, irreversible blindness, vertigo, and hypertension, following edema of the face, throat, neck, tongue, ingestion, and larynx coupled with respiratory distress. Aromatic amines may be made mobile by water or perspiration, which makes it easier for the body to absorb them through the skin and other exposed regions, such as the mouth. Ingestion is a more rapid absorption method, making it potentially more harmful because a greater quantity of dye may be absorbed in less time [126]. When digested by the enzymes in the liver, water-soluble azo dyes can produce toxic byproducts.

More than one hundred of these dyes that have the potential for the creation of carcinogenic amines are still accessible on the market, even though agreements have been made to ban the thousands of dyes whose toxicity has been researched. Export needs coupled with the availability of low-cost labor are responsible for creating small-scale textile manufacturers in several parts of the world, such as India [127]. These facilities are responsible for the covert discharge of poisonous colors into bodies of water.

References

[1] Benkhaya S, M' rabet S and El Harfi A 2020 A review on classifications, recent synthesis and applications of textile dyes *Inorg. Chem. Commun.* **115** 107891

[2] Abd-Elhamid A I, Emran M, El-Sadek M H, El-Shanshory A A, Soliman H M A, Akl M A and Rashad M 2020 Enhanced removal of cationic dye by eco-friendly activated biochar derived from rice straw *Appl. Water Sci.* **10** 45

[3] Bouras H D, Isik Z, Arikan E B, Yeddou A R, Bouras N, Chergui A, Favier L, Amrane A and Dizge N 2021 Biosorption characteristics of methylene blue dye by two fungal biomasses *Int. J. Environ. Stud.* **78** 365–81

[4] Tara N, Siddiqui S I, Rathi G, Chaudhry S A, Inamuddin A M and Asiri 2019 Nano-engineered adsorbent for the removal of dyes from water: a review *Curr. Anal. Chem.* **16** 14–40

[5] Khan I, Khan I, Usman M, Imran M and Saeed K 2020 Nanoclay-mediated photocatalytic activity enhancement of copper oxide nanoparticles for enhanced methyl orange photo-degradation *J. Mater. Sci., Mater. Electron.* **31** 8971–85

[6] Ahmad M, Rehman W, Khan M M, Qureshi M T, Gul A, Haq S, Ullah R, Rab A and Menaa F 2021 Phytogenic fabrication of ZnO and gold decorated ZnO nanoparticles for photocatalytic degradation of rhodamine B *J. Environ. Chem. Eng.* **9** 104725

[7] Pandey S, Do J Y, Kim J and Kang M 2020 Fast and highly efficient removal of dye from aqueous solution using natural locust bean gum based hydrogels as adsorbent *Int. J. Biol. Macromol.* **143** 60–75

[8] Fong W M, Affam A C and Chung W C 2020 Synthesis of Ag/Fe/CAC for colour and COD removal from methylene blue dye wastewater *Int. J. Environ. Sci. Technol.* **17** 3485–94

[9] Hanafi M F and Sapawe N 2020 Effect of pH on the photocatalytic degradation of Remazol Brilliant Blue dye using zirconia catalyst *Mater. Today Proc.* **31** 260–2

[10] Liu L, Liu S, Mishra S B and Sheng L 2019 Iron-included mesoporous silica thin slice as adsorbent and Fenton-like catalyst for the adsorption and degradation of dye in wastewater *Ceram. Int.* **45** 15475–81

[11] Gupta V K and Suhas 2009 Application of low-cost adsorbents for dye removal—a review *J. Environ. Manage.* **90** 2313–42

[12] Demirbas A 2009 Agricultural based activated carbons for the removal of dyes from aqueous solutions: a review *J. Hazard. Mater.* **167** 1–9

[13] Forgacs E, Cserháti T and Oros G 2004 Removal of synthetic dyes from wastewaters: a review *Environ. Int.* **30** 953–71

[14] Ali M and Sreekrishnan T R 2001 Aquatic toxicity from pulp and paper mill effluents: a review *Adv. Environ. Res.* **5** 175–96

[15] Haji Azaman S A, Afandi A, Hameed B H and Mohd Din A T 2018 Removal of malachite green from aqueous phase using coconut shell activated carbon: adsorption, desorption, and reusability studies *J. Appl. Sci. Eng.* **21** 317–30

[16] Sriram G, Kigga M, Uthappa U T, Rego R M, Thendral V, Kumeria T, Jung H Y and Kurkuri M D 2020 Naturally available diatomite and their surface modification for the removal of hazardous dye and metal ions: a review *Adv. Colloid Interface Sci.* **282** 102198

[17] Banerjee S, Sharma G C, Gautam R K, Chattopadhyaya M C, Upadhyay S N and Sharma Y C 2016 Removal of malachite green, a hazardous dye from aqueous solutions using *Avena sativa* (oat) hull as a potential adsorbent *J. Mol. Liq.* **213** 162–72

[18] Raval N P, Shah P U and Shah N K 2017 Malachite green 'a cationic dye' and its removal from aqueous solution by adsorption *Appl. Water Sci.* **7** 3407–45

[19] Reis H C O, Cossolin A S, Santos B A P, Castro K C, Pereira G M, Silva V C, Sousa P T Jr, Dall'Oglio E L, Vasconcelos L G and Morais E B 2018 Malt bagasse waste as biosorbent for malachite green: an ecofriendly approach for dye removal from aqueous solution *Int. J. Biol. Life Agric. Sci.* **12** 118–27

[20] Kumar Srivastav A, Roy D and Ashvani Kumar Srivastav C 2015 Effects of malachite green (triarylmethane dye) and Pyceze (bronopol) on the hematological parameters of a freshwater catfish *Heteropneustes fossilis* (Bloch) *Int. J. Fish. Aquat. Stud.* **2** 119–22

[21] Yonar M E and Yonar S M 2010 Changes in selected immunological parameters and antioxidant status of rainbow trout exposed to malachite green (*Oncorhynchus mykiss*, Walbaum, 1792) *Pestic. Biochem. Physiol.* **97** 19–23

[22] Tongpoothorn W, Somsimee O, Somboon T and Sriuttha M 2019 An alternative and cost-effective biosorbent derived from napier grass stem for malachite green removal *J. Mater. Environ. Sci.* **10** 685–95

[23] Khan I *et al* 2022 Review on methylene blue: its properties, uses, toxicity and photodegradation *Water* **14** 242

[24] Yang C, Dong W, Cui G, Zhao Y, Shi X, Xia X, Tang B and Wang W 2017 Highly efficient photocatalytic degradation of methylene blue by P2ABSA-modified TiO_2 nanocomposite due to the photosensitization synergetic effect of TiO_2 and P2ABSA *RSC Adv.* **7** 23699–708

[25] Koyuncu H and Kul A R 2020 Removal of methylene blue dye from aqueous solution by nonliving lichen (*Pseudevernia furfuracea* (L.) Zopf.), as a novel biosorbent *Appl. Water Sci.* **10** 72

[26] Manimohan M, Pugalmani S, Ravichandran K and Sithique M A 2020 Synthesis and characterisation of novel Cu(II)-anchored biopolymer complexes as reusable materials for the photocatalytic degradation of methylene blue *RSC Adv.* **10** 18259–79

[27] Contreras M, Grande-Tovar C D, Vallejo W and Chaves-López C 2019 Bio-removal of methylene blue from aqueous solution by *Galactomyces geotrichum* KL20A *Water* **11** 282

[28] Millicent Mabel M, Sundararaman T R, Parthasarathy N and Rajkumar J 2019 Chitin beads from *Peneaus* sp. shells as a biosorbent for methylene blue dye removal *Polish J. Environ. Stud.* **28** 2253–9

[29] Ebi Ebi O, Falilat Taiwo A and Tunde Folorunsho A 2018 Kinetic modelling of the biosorption of methylene blue onto wild melon (*Lagenaria sphaerica*) *Am. J. Chem. Eng.* **6** 126–34

[30] Lawagon C P and Amon R E C 2020 Magnetic rice husk ash 'cleanser' as efficient methylene blue adsorbent *Environ. Eng. Res.* **25** 685–92

[31] Faizal A N M, Putra N R and Zaini M A A 2022 Insight into the adsorptive mechanisms of methyl violet and reactive orange from water—a short review *Part. Sci. Technol.* **0** 1–10

[32] Üner O, Geçgel Ü, Kolancilar H and Bayrak Y 2017 Adsorptive removal of rhodamine B with activated carbon obtained from okra wastes *Chem. Eng. Commun.* **204** 772–83

[33] Kooh M R R, Lim L B L, Lim L H and Dahri M K 2016 Separation of toxic rhodamine B from aqueous solution using an efficient low-cost material, *Azolla pinnata*, by adsorption method *Environ. Monit. Assess.* **188** 1–15

[34] Tariq M, Muhammad M, Khan J, Raziq A, Uddin M K, Niaz A, Ahmed S S and Rahim A 2020 Removal of rhodamine B dye from aqueous solutions using photo-Fenton processes and novel Ni–Cu@MWCNTs photocatalyst *J. Mol. Liq.* **312** 113399

[35] Soejima T, Nishizawa K and Isoda R 2018 Monodisperse manganese oxide nanoparticles: synthesis, characterization, and chemical reactivity *J. Colloid Interface Sci.* **510** 272–9

[36] Hamdaoui O 2011 Intensification of the sorption of rhodamine B from aqueous phase by loquat seeds using ultrasound *Desalination* **271** 279–86

[37] Imam S S and Babamale H F 2020 A short review on the removal of rhodamine B dye using agricultural waste-based adsorbents *Asian J. Chem. Sci.* **7** 25–37

[38] Bhat S A, Rashid N, Rather M A, Bhat S A, Ingole P P and Bhat M A 2020 Highly efficient catalytic reductive degradation of rhodamine-B over palladium-reduced graphene oxide nanocomposite *Chem. Phys. Lett.* **754** 137724

[39] Mani S and Bharagava R N 2016 Exposure to crystal violet, its toxic, genotoxic and carcinogenic effects on environment and its degradation and detoxification for environmental safety *Rev. Environ. Contam. Toxicol.* **237** 71–104

[40] Wang S, Boyjoo Y, Choueib A and Zhu Z H 2005 Removal of dyes from aqueous solution using fly ash and red mud *Water Res.* **39** 129–38

[41] Abbas M, Harrache Z and Trari M 2019 Removal of gentian violet in aqueous solution by activated carbon equilibrium, kinetics, and thermodynamic study *Adsorpt. Sci. Technol.* **37** 566–89

[42] Doke K M, Yusufi M, Joseph R D and Khan E M 2016 Comparative adsorption of crystal violet and Congo red onto $ZnCl_2$ activated carbon *J. Dispers. Sci. Technol.* **37** 1671–81

[43] Chakraborty S, Chowdhury S and Das Saha P 2011 Adsorption of crystal violet from aqueous solution onto NaOH-modified rice husk *Carbohydr. Polym.* **86** 1533–41

[44] Khan M M R, Rahman M W, Ong H R, Ismail A B and Cheng C K 2016 Tea dust as a potential low-cost adsorbent for the removal of crystal violet from aqueous solution *Desalin. Water Treat.* **57** 14728–38

[45] Ahmad R 2009 Studies on adsorption of crystal violet dye from aqueous solution onto coniferous pinus bark powder (CPBP) *J. Hazard. Mater.* **171** 767–73

[46] Rehman R, Mahmud T and Irum M 2015 Brilliant green dye elimination from water using *Psidium guajava* leaves and *Solanum tuberosum* peels as adsorbents in environmentally benign way *J. Chem.* **2015** 126036

[47] Nandi B K, Goswami A and Purkait M K 2009 Adsorption characteristics of brilliant green dye on kaolin *J. Hazard. Mater.* **161** 387–95

[48] Pohanish R P 2017 *Sittig's Handbook of Toxic and Hazardous Chemicals and Carcinogens* 7th edn (Amsterdam: Elsevier Science)

[49] Roy K, Verma K M, Vikrant K, Goswami M, Sonwani R K, Rai B N, Vellingiri K, Kim K H, Giri B S and Singh R S 2018 Removal of Patent Blue (V) dye using Indian bael shell biochar: characterization application and kinetic studies *Sustainability* **10** 2669

[50] Ibrahim Y and Gaya U I 2020 Comparative optimization of removal of low levels of brilliant green by ZnO photocatalysis and photo-Fenton *J. Mater. Environ. Sci.* **11** 318–32

[51] Mansour R A, El Shahawy A, Attia A and Beheary M S 2020 Brilliant green dye biosorption using activated carbon derived from guava tree wood *Int. J. Chem. Eng.* **2020** 8053828

[52] Zeyada H M, Makhlouf M M, Ismail M I M and Salama A A 2015 Thermal behavior, structure formation and optical characteristics of nanostructured basic fuchsine thin films *Mater. Chem. Phys.* **163** 45–53

[53] Ong S T, Tan S Y, Khoo E C, Lee S L and Ha S T 2012 Equilibrium studies for basic blue 3 adsorption onto durian peel (*Durio zibethinus* Murray) *Desalin. Water Treat.* **45** 161–9

[54] Qin J, Qiu F, Rong X, Yan J, Zhao H and Yang D 2014 Removal of basic fuchsin dye from aqueous solutions using graphite oxide modified aromatic polyurethane foam material *Toxicol. Environ. Chem.* **96** 849–60

[55] Loera-Serna S, Ortiz E and Beltrán H I 2017 First trial and physicochemical studies on the loading of basic fuchsin, crystal violet and Black Eriochrome T on HKUST-1 *New J. Chem.* **41** 3097–105

[56] Ghoniem M G, Ali F A M, Abdulkhair B Y, Elamin M R A, Alqahtani A M, Rahali S and Ben Aissa M A 2022 Highly selective removal of cationic dyes from wastewater by MgO nanorods *Nanomaterials* **12** 1023
[57] Yang X, Li Y, Du Q, Sun J, Chen L, Hu S, Wang Z, Xia Y and Xia L 2015 Highly effective removal of basic fuchsin from aqueous solutions by anionic polyacrylamide/graphene oxide aerogels *J. Colloid Interface Sci.* **453** 107–14
[58] Haddad M E 2016 Removal of basic fuchsin dye from water using mussel shell biomass waste as an adsorbent: equilibrium, kinetics, and thermodynamics *J. Taibah Univ. Sci.* **10** 664–74
[59] Wu L, Liu X, Lv G, Zhu R, Tian L, Liu M, Li Y, Rao W, Liu T and Liao L 2021 Study on the adsorption properties of methyl orange by natural one-dimensional nano-mineral materials with different structures *Sci. Rep.* **11** 1–11
[60] ProtonsTalk n.d. *Methyl Orange—Structure, Formula, Properties, Preparation* https://protonstalk.com/acids-bases-salts/methyl-orange/ (Accessed: 23 November 2022)
[61] Maruthanayagam A, Mani P, Kaliappan K and Chinnappan S 2020 *In vitro* and *in silico* studies on the removal of methyl orange from aqueous solution using *Oedogonium subplagiostomum* AP1 *Water Air Soil Pollut.* **231** 1–21
[62] Farghali M A, Selim A M, Khater H F, Bagato N, Alharbi W, Alharbi K H and Taha Radwan I 2022 Optimized adsorption and effective disposal of Congo red dye from wastewater: hydrothermal fabrication of MgAl-LDH nanohydrotalcite-like materials *Arab. J. Chem.* **15** 104171
[63] Priyadarshini B, Patra T and Sahoo T R 2021 An efficient and comparative adsorption of Congo red and trypan blue dyes on MgO nanoparticles: kinetics, thermodynamics and isotherm studies *J. Magnes. Alloy* **9** 478–88
[64] Harja M, Buema G and Bucur D 2022 Recent advances in removal of Congo red dye by adsorption using an industrial waste *Sci. Rep.* **12** 1–18
[65] Alamrani N A and Al-Aoh H A 2021 Elimination of Congo red dye from industrial wastewater using *Teucrium polium* L. as a low-cost local adsorbent *Adsorpt. Sci. Technol.* **2021** 5728696
[66] Litefti K, Freire M S, Stitou M and González-Álvarez J 2019 Adsorption of an anionic dye (Congo red) from aqueous solutions by pine bark *Sci. Rep.* **9** 1–11
[67] Zhou Y, Ge L, Fan N and Xia M 2018 Adsorption of Congo red from aqueous solution onto shrimp shell powder *Adsorpt. Sci. Technol.* **36** 1310–30
[68] Machado F M, Carmalin S A, Lima E C, Dias S L P, Prola L D T, Saucier C, Jauris I M, Zanella I and Fagan S B 2016 Adsorption of Alizarin Red S dye by carbon nanotubes: an experimental and theoretical investigation *J. Phys. Chem.* C **120** 18296–306
[69] Rehman R and Mahmud T 2013 Sorptive elimination of Alizarin Red-S dye from water using citrullus lanatus peels in environmentally benign way along with equilibrium data modeling *Asian J. Chem.* **25** 5351–6
[70] Singh P and Naskar N 2021 Fabrication of In(III)–Alizarin Red S complex trap for efficient detection of fluoride ion in aqueous environs *J. Anal. Sci. Technol.* **12** 1–10
[71] Albadarin A B and Mangwandi C 2015 Mechanisms of Alizarin Red S and methylene blue biosorption onto olive stone by-product: isotherm study in single and binary systems *J. Environ. Manage.* **164** 86–93
[72] Hossain A, Rayhan A B M S, Raihan M J, Nargis A, Ismail I M I, Habib A and Mahmood A J 2016 Kinetics of degradation of eosin Y by one of the advanced oxidation processes (AOPs)—Fenton's process *Am. J. Anal. Chem.* **7** 863–79

[73] Abd El-Gelil G M, Mansour M S, Ebrahiem E E, EL-Shazly A H and Abou-Gabal H 2020 Degradation of eosin Y in water by corona treatment with a dielectric barrier discharge plasma *Chem. Eng. Technol.* **43** 2015–22

[74] Majek M, Filace F and Wangelin A J V 2014 On the mechanism of photocatalytic reactions with eosin Y *Beilstein J. Org. Chem.* **10** 981–9

[75] Assefi P, Ghaedi M, Ansari A, Habibi M H and Momeni M S 2014 Artificial neural network optimization for removal of hazardous dye eosin Y from aqueous solution using Co_2O_3-NP-AC: isotherm and kinetics study *J. Ind. Eng. Chem.* **20** 2905–13

[76] Jedynak K, Repelewicz M, Kurdziel K and Wideł D 2020 Removal of orange II from aqueous solutions using micro-mesoporous carbon materials: kinetic and equilibrium studies *Desalin. Water Treat.* **190** 294–311

[77] Riaz N, Chong F K, Dutta B K, Man Z B, Khan M S and Nurlaela E 2012 Photodegradation of orange II under visible light using Cu–Ni/TiO_2: effect of calcination temperature *Chem. Eng. J.* **185–86** 108–19

[78] Gupta V K, Mittal A, Gajbe V and Mittal J 2006 Removal and recovery of the hazardous azo dye acid orange 7 through adsorption over waste materials: bottom ash and de-oiled soya *Ind. Eng. Chem. Res.* **45** 1446–53

[79] Gupta S V and Ahmaruzzaman M 2022 $CeO_2/Fe_3O_4/g\text{-}C_3N_4$ nanohybrid for adsorptive removal of rose bengal from aqueous stream *Int. J. Environ. Anal. Chem.*

[80] Murugadoss G, Kumar D D, Kumar M R, Venkatesh N and Sakthivel P 2021 Silver decorated CeO_2 nanoparticles for rapid photocatalytic degradation of textile rose bengal dye *Sci. Rep.* **11** 1–13

[81] Zhou M, Zhang Y, Chen Y, Zhang F and Yang D 2022 Optimization of the decolorization conditions of rose bengal by using *Aspergillus niger* TF05 and a decolorization mechanism *Microbiology* **168** 001128

[82] Yoguim M I, Grandini G S, de L, Bertozo C, Caracelli I, Ximenes V F and de Souza A R 2022 Studies on the interaction of rose bengal with the human serum albumin protein under spectroscopic and docking simulations aspects in the characterization of binding sites *Chemosensors* **10** 440

[83] Wang S, Wang M, Liu Y, Hu D, Gu L, Fei X and Zhang J 2019 Effect of rapamycin microspheres in Sjögren syndrome dry eye: preparation and outcomes *Ocul. Immunol. Inflamm.* **27** 1357–64

[84] Gupta V K, Mittal A, Jhare D and Mittal J 2012 Batch and bulk removal of hazardous colouring agent rose bengal by adsorption techniques using bottom ash as adsorbent *RSC Adv.* **2** 8381–9

[85] Zubair M, Aziz H A, Ihsanullah I, Ahmad M A and Al-Harthi M A 2022 Enhanced removal of Eriochrome Black T from water using biochar/layered double hydroxide/chitosan hybrid composite: performance evaluation and optimization using BBD-RSM approach *Environ. Res.* **209** 112861

[86] Alp H, Ince M, Ince O K and Onal A 2020 Biosorptive removal of Eriochrome Black-T using *Agaricus campestris*: parameters optimization with response surface methodology *Desalin. Water Treat.* **175** 244–54

[87] Altaher H, Khalil T E and Abubeah R 2014 The effect of dye chemical structure on adsorption on activated carbon: a comparative study *Color. Technol.* **130** 205–14

[88] Moeinpour F, Alimoradi A and Kazemi M 2014 Efficient removal of Eriochrome Black-T from aqueous solution using $NiFe_2O_4$ magnetic nanoparticles *J. Environ. Heal. Sci. Eng.* **12** 1–7

[89] Abbas R F, Hami H K, Mahdi N I and Waheb A A 2020 Removal of Eriochrome Black T dye by using Al_2O_3 nanoparticles: central composite design, isotherm and error analysis *Iran. J. Sci. Technol. Trans.* A **44** 993–1000

[90] Kulkarni M R, Bhagyalakshmi C, Anand D and Herle R N 2018 Removal of Remazol Brilliant Blue dye from aqueous solutions using water hyacinth root powder *Desalin. Water Treat.* **122** 331–8

[91] Hadibarata T and Kristanti R A 2012 Effect of environmental factors in the decolorization of Remazol Brilliant Blue R by polyporus SP. S133 *J. Chil. Chem. Soc.* **57** 1095–8

[92] Arslantaş C, M'barek I, Saleh M, Isik Z, Ozdemir S, Dundar A and Dizge N 2022 Basic Red 18 and Remazol Brilliant Blue R biosorption using *Russula brevipes, Agaricus augustus, Fomes fomentarius Water Pract. Technol.* **17** 749–62

[93] De Keijzer M, Van Bommel M R, Hofmann-De Keijzer R, Knaller R and Oberhumer E 2012 Indigo carmine: understanding a problematic blue dye *Stud. Conserv.* **57** S87–S95

[94] Zhang J, Zhou Q and Ou L 2016 Removal of indigo carmine from aqueous solution by microwave-treated activated carbon from peanut shell *Desalin. Water Treat.* **57** 718–27

[95] Volkov V V, Chelli R, Righini R and Perry C C 2020 Indigo chromophores and pigments: structure and dynamics *Dyes Pigm.* **172** 107761

[96] Chowdhury M F, Khandaker S, Sarker F, Islam A, Rahman M T and Awual M R 2020 Current treatment technologies and mechanisms for removal of indigo carmine dyes from wastewater: a review *J. Mol. Liq.* **318** 114061

[97] Khandaker S, Toyohara Y, Kamida S and Kuba T 2018 Adsorptive removal of cesium from aqueous solution using oxidized bamboo charcoal *Water Resour. Ind.* **19** 35–46

[98] Turcanu A and Bechtold T 2011 Indirect cathodic reduction of dispersed indigo by 1,2-dihydroxy-9,10-anthraquinone-3-sulphonate (Alizarin Red S) *J. Solid State Electrochem.* **15** 1875–84

[99] Ramig K *et al* 2015 The nature of thermochromic effects in dyeings with indigo, 6-bromoindigo, and 6,6′-dibromoindigo, components of Tyrian purple *Dyes Pigm.* **117** 37–48

[100] Trujillo-Reyes J, Sánchez-Mendieta V, Colín-Cruz A and Morales-Luckie R A 2010 Removal of indigo blue in aqueous solution using Fe/Cu nanoparticles and C/Fe–Cu nanoalloy composites *Water Air Soil Pollut.* **207** 307–17

[101] Rauh-Hain J A and Laufer M R 2011 Increased diagnostic accuracy of laparoscopy in endometriosis using indigo carmine: a new technique *Fertil. Steril.* **95** 1113–4

[102] Paul R, Blackburn R S and Bechtold T 2021 Indigo and indigo colorants *Ullmann's Encyclopedia of Industrial Chemistry* (New York: Wiley) pp 1–16

[103] Zhang J, Zhang P, Zhang S and Zhou Q 2014 Comparative study on the adsorption of tartrazine and indigo carmine onto maize cob carbon *Sep. Sci. Technol.* **49** 877–86

[104] Katheresan V, Kansedo J and Lau S Y 2018 Efficiency of various recent wastewater dye removal methods: a review *J. Environ. Chem. Eng.* **6** 4676–97

[105] Hunger K, Gregory P, Miederer P, Berneth H, Heid C and Mennicke W 2004 Important chemical chromophores of dye classes *Industrial Dyes: Chemistry, Properties, Applications* (New York: Wiley) pp 13–112

[106] Eren E and Afsin B 2007 Investigation of a basic dye adsorption from aqueous solution onto raw and pre-treated sepiolite surfaces *Dyes Pigm.* **73** 162–7

[107] Yunusa U, Usman B and Bashir Ibrahim M 2021 Cationic dyes removal from wastewater by adsorptive method: a systematic in-depth review *Alger. J. Chem. Eng.* **2** 6–40

[108] Marungrueng K and Pavasant P 2007 High performance biosorbent (*Caulerpa lentillifera*) for basic dye removal *Bioresour. Technol.* **98** 1567–72

[109] Chan L S, Cheung W H, Allen S J and McKay G 2017 Equilibrium adsorption isotherm study of binary basic dyes on to bamboo derived activated carbon *HKIE Trans. Hong Kong Inst. Eng.* **24** 182–92

[110] Eren E 2009 Investigation of a basic dye removal from aqueous solution onto chemically modified Unye bentonite *J. Hazard. Mater.* **166** 88–93

[111] Mao J, Won S W, Min J and Yun Y S 2008 Removal of Basic Blue 3 from aqueous solution by *Corynebacterium glutamicum* biomass: biosorption and precipitation mechanisms *Korean J. Chem. Eng.* **25** 1060–4

[112] Sriram G, Kigga M, Uthappa U T, Rego R M, Thendral V, Kumeria T, Jung H Y and Kurkuri M D 2020 Naturally available diatomite and their surface modification for the removal of hazardous dye and metal ions: a review *Adv. Colloid Interface Sci.* **282** 102198

[113] Hunger K 2004 Dyes, general survey *Industrial Dyes: Chemistry, Properties, Applications* (New York: Wiley) ch 1 pp 1–12

[114] Tabak A, Eren E, Afsin B and Caglar B 2009 Determination of adsorptive properties of a Turkish sepiolite for removal of Reactive Blue 15 anionic dye from aqueous solutions *J. Hazard. Mater.* **161** 1087–94

[115] Attia A A, Rashwan W E and Khedr S A 2006 Capacity of activated carbon in the removal of acid dyes subsequent to its thermal treatment *Dyes Pigm.* **69** 128–36

[116] Hassaan M A and El Nemr A 2017 Health and environmental impacts of dyes: mini review *Am. J. Environ. Sci. Eng.* **1** 64–7

[117] Hassaan M A and El Nemr A 2017 Advanced oxidation processes for textile wastewater treatment *Int. J. Photochem. Photobiol.* **2** 85–93

[118] Al-Kdasi A, Idris A, Saed K and Guan C T 2004 Treatment of textile wastewater by advanced oxidation processes—a review *Glob. Nest. J.* **6** 222–30

[119] Hassaan M A, El Nemr A and Madkour F F 2016 CODEN(USA): PCJHBA application of ozonation and UV assisted ozonation for decolorization of direct yellow 50 in sea water *Pharm. Chem. J.* **3** 131–8

[120] Tüfekci N, Sivri N and Toroz İ 2007 Pollutants of textile industry wastewater and assessment of its discharge limits by water quality standards *Turkish J. Fish. Aquat. Sci.* **7** 97–103

[121] Venkata Mohan S, Chandrasekhar Rao N and Karthikeyan J 2002 Adsorptive removal of direct azo dye from aqueous phase onto coal based sorbents: a kinetic and mechanistic study *J. Hazard. Mater.* **90** 189–204

[122] Hassaan M A, El Nemr A and Madkour F F 2017 Testing the advanced oxidation processes on the degradation of Direct Blue 86 dye in wastewater *Egypt. J. Aquat. Res.* **43** 11–9

[123] Savin I I and Butnaru R 2008 Wastewater characteristics in textile finishing mills *Environ. Eng. Manag. J.* **7** 859–64

[124] Chung K T 1983 The significance of azo-reduction in the mutagenesis and carcinogenesis of azo dyes *Mutat. Res. Genet. Toxicol.* **114** 269–81

[125] Apostol L C, Pereira L, Pereira R, Gavrilescu M and Alves M M 2012 Biological decolorization of xanthene dyes by anaerobic granular biomass *Biodegradation* **23** 725–37

[126] Puvaneswari N, Muthukrishnan J and Gunasekaran P 2006 Toxicity assessment and microbial degradation of azo dyes *Indian J. Exp. Biol.* **44** 618–26

[127] Lellis B, Fávaro-Polonio C Z, Pamphile J A and Polonio J C 2019 Effects of textile dyes on health and the environment and bioremediation potential of living organisms *Biotechnol. Res. Innov.* **3** 275–90

IOP Publishing

Photocatalytic Dye Degradation Using Green Polymeric-based Nanostructures
Principles and applications in wastewater treatment

Muhammad Ikram, Muhammad Ahsaan Bari and Junaid Haider

Chapter 4

Structural and morphological characteristics of green polymers

Green polymers, also known as biopolymers, are natural compounds that are extracted components from plants and other living things, and are made up of chain-like biomolecules. They are basic chemical building blocks that are obtained from renewable resources and have the potential to degrade in the environment. It is becoming increasingly common to use biomaterials to cut down on the consumption of non-renewable resources and reduce the pollution in the environment caused by the production of synthetic materials. Due to the fact that they possess exceptional physical and chemical properties, polysaccharide polymers are regarded as being of significant importance. Cellulose, chitin, chitosan, starch, and alginate are some examples of polysaccharide polymers. These are regarded as being excellent materials for the degradation of organic dyes, in addition to a wide variety of additional uses that are beneficial to the environment. In this chapter the most well-known polysaccharides are examined together with their structure and chemistry in relation to the degradation of organic dyes. Lastly, various structural and morphological characterization methods for green polymers are discussed, such as scanning electron microscopy (SEM), transmission electron microscopy (TEM), x-ray diffraction (XRD), x-ray photoelectron spectroscopy (XPS), Raman spectroscopy, and Fourier transform infrared spectroscopy (FTIR).

4.1 Biopolymers

Biopolymers are natural polymeric molecules that are produced by living organisms. They are made up of monomer units covalently bound together to produce bigger molecules with a high molecular weight. The three most important biopolymers are polynucleotides, polypeptides, and polysaccharides. DNA and RNA are polynucleotides, constructed from nucleotide monomers, whereas polypeptides are

doi:10.1088/978-0-7503-5517-9ch4

polymers that use amino acids as their monomer unit. Some examples of polypeptides are albumin, fibrin, zein, collagen, and keratin. Polysaccharides have the simplest structure compared to the other two types of biopolymers. These molecules are linear or branching polymeric carbohydrate chains. Some examples of polysaccharides include cellulose, chitin, chitosan, starch, and alginate [1]. Insects, green algae, mammalian bone, and fungi are some examples of the natural sources of some of the most important biopolymers, as depicted in figure 4.1.

4.2 Production of biopolymeric nanomaterials

The conventional approaches to producing polymeric nanomaterials include the utilization of chemical compounds that, due to their inherent stability over time, have the potential to subsequently result in environmental toxicity. In contrast, biopolymers are often made up of non-hazardous monomers and have no impact on the atmosphere's carbon content. Because the shape of the biomass is frequently organized at the nano-level, the production of nanomaterials is made easier through the use of biopolymers. Several different kinds of living things, including plants, algae, fungi, bacteria, and mammals, have biopolymers in their tissues. Starch, alginate, chitosan, dextran, and chitin are all classified as macromolecules (figure 4.2). These biopolymers are frequently used in the production of nanomaterials [3–5]. Cellulose, dextran, starch, and chitosan are polysaccharides generated from microbial biofilms and plants. Chitosan, in particular, has been utilized for the elimination of dye as well as the treatment of wastewater [6, 7]. The chemical make-

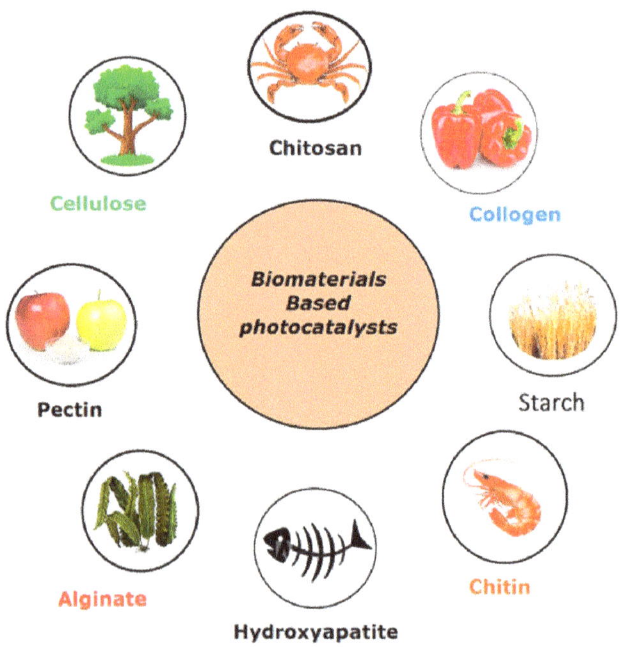

Figure 4.1. Natural sources of some commonly used biopolymers in photocatalysis. (Reproduced with permission from [2]. Copyright 2021 Elsevier.)

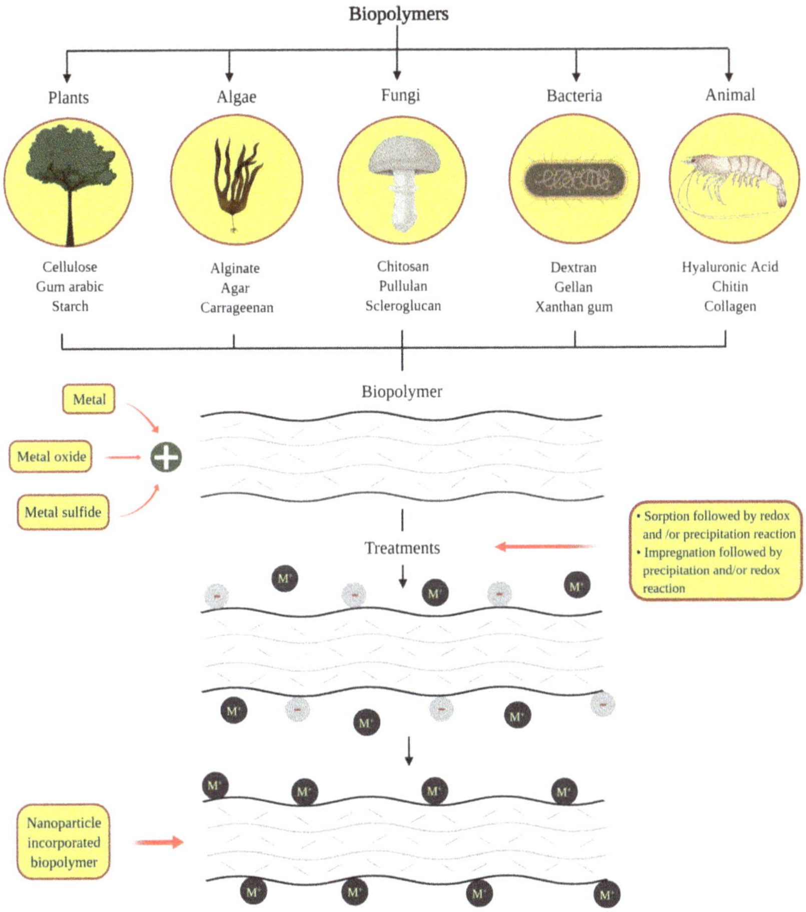

Figure 4.2. Different sources of biopolymers and their use in the synthesis of green polymeric nanomaterials. M^+: metal ions. (Reproduced with permission from [12]. Copyright 2020 Springer Nature.)

up of biopolymers is one of a kind, exhibiting a wide range of physiological characteristics. Biopolymers can be combined with metals to provide the ideal environment for synthesizing biopolymeric nanomaterials. The intermolecular hydrogen bonding between biopolymers, in particular, is responsible for the development of molecular capsules. For instance, starch has the potential to integrate metals or metal oxides, which results in the formation of polymeric nanocomposites. Chitosan, owing to its wide adaptability, is also put to use in applications that are connected to nanotechnology [8, 9]. The formation of nanomaterials is possible through the supramolecular incorporation of silver (Ag) inside starch [10]. The processes of sorption and impregnation are both viable options for nanomaterials incorporation into biopolymers [11]. Solid colloidal particles known as polymeric nanomaterials have a size range of 10 nm to 1 μm.

4.3 Polysaccharide-derived nanocatalysts for water treatment

Polysaccharides are macromolecules that are produced from multiple monosaccharide units and are then bonded together by glycosidic connections. Polysaccharides have a variety of distinct biological roles that range from cell signaling to immunological recognition. Polysaccharide-derived nanocatalysts have recently been the subject of research as potential heterogeneous and unique catalysts for water treatment. These nanocatalysts have shown great catalytic ability. In this section, a summary of the different polysaccharides, along with their morphological and structural characteristics, is provided.

4.3.1 Cellulose: chemistry and properties

In addition to being one of the most common substances on Earth, cellulose is the primary constituent of all plant fibers and is responsible for the tensile strength of plants. Cellulose is one of the most prevalent compounds in the Universe [13]. In addition, due to its availability, biodegradability, renewability, and inexpensiveness, it is the most promising biopolymer currently available. The molecular formula for cellulose is $(C_6H_{10}O_5)_n$ and it is a linear polysaccharide with long chains. The main cellulose component is a dimer called cellobiose, which is made up of β-D-glucopyranose units connected by β-1,4-glycosidic linkages. Plant-based products such as paper, wood, and grass contain cellulose (figure 4.3).

Cellulose may be extracted from a wide variety of lignocellulosic materials, such as grasses (bamboo, bagasse), annual plants (ramie, hemp, flax, jute), and wood (both softwoods and hardwoods), but it is also possible to obtain cellulose from other sources, such as bacterial or animal material. Scientists have been able to study the nanoscale structure of cellulose in recent years thanks to advancements in production technology and analytical instrumentation, which, in comparison to its microscale and macroscopic counterparts, reveals significant differences in their properties. Compared to natural cellulose, cellulose nanoparticles, which range in diameter from a few nanometers to tens of nanometers, offer better mechanical properties. These enhancements result from the removal of macroscopic defects that are built into the microfibers. As shown in figure 4.3, cellulose nanoparticles offer a number of benefits to their users.

Nanocellulose can be divided into two major categories, according to the Technical Association of the Pulp and Paper Industry (TAPPI) and the USDA Forest Service: cellulose nanocrystals (CNCs), also known as cellulose whiskers, and cellulose nanofibrils (CNF), also known as micro fibrillated cellulose [14]. Nanocellulose is also thought to include electrospun cellulose nanofibers (ECNF) and bacterial cellulose (BC). The most common cellulosic nanomaterials, CNF and CNC, are created by the top-down method of cellulose fiber breakdown into nanoscale particles, whereas ECNF and BC are created by the bottom-up method of nanofiber accumulation from low molecular weight sugars produced by bacteria or from cellulose derivatives produced by electrospinning, respectively. Table 4.1 shows the properties of cellulose based nanomaterials. The process of producing cellulose nanoparticles is outlined and illustrated in the following sections.

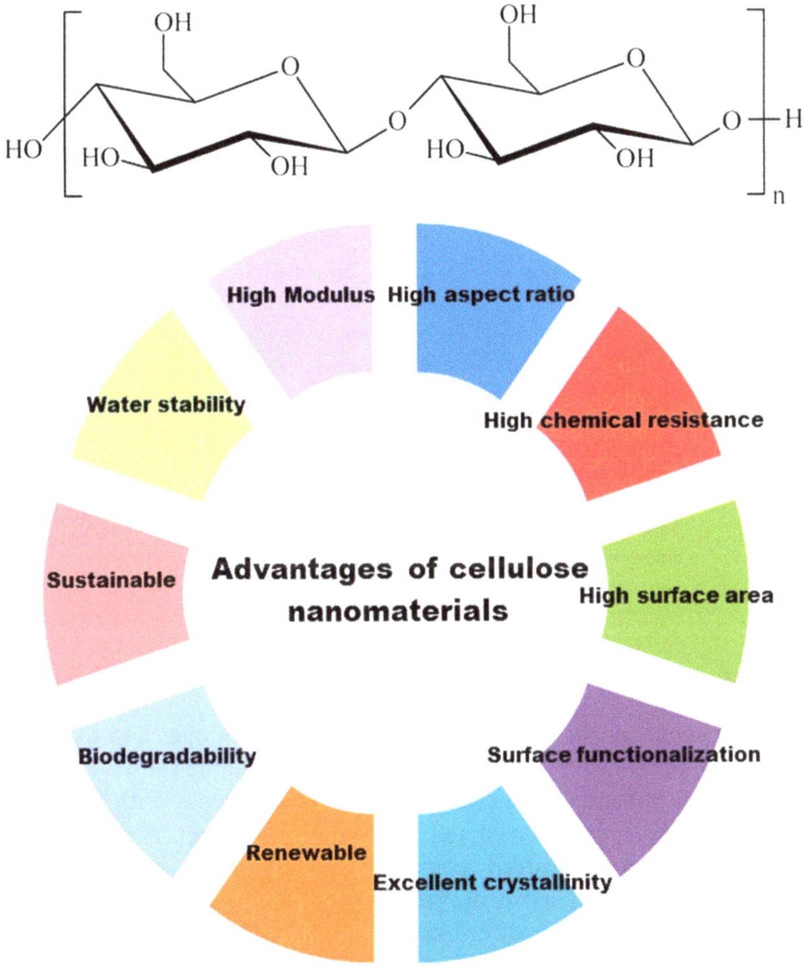

Figure 4.3. Basic chemical structure of cellulose and some advantages of cellulose nanomaterials. (Reproduced with permission from [15]. Copyright 2019 American Chemical Society.)

4.3.1.1 Cellulose nanocrystals

The process for producing cellulose nanocrystals (CNCs) includes acid hydrolysis of cellulose fibers. To create clearly defined crystalline domains, this procedure breaks down amorphous cellulose fibers. The amorphous cellulose is broken down into the solution by acid hydrolysis, allowing CNC to be extracted from the cellulose [16]. Consequently, a crystalline structure known as nanocrystals of cellulose is produced [17]. These environmentally friendly nanomaterials have a high level of functionality, high biocompatibility of the surface, robust mechanical tolerance, and larger specific surface area. They are heavier than steel and have a more compact profile than aluminum. Depending on their origin, they also have a huge aspect ratio (length/diameter), ranging from 30 to 150 [18]. In addition to its maximum

Table 4.1. Characteristics of nanomaterials based on cellulose [27].

Cellulose nanomaterials	CNCs	CNFs	BNC
Source	Cotton fibers, lotus leaves, rice husk, wood, flax, etc	Sugarcane bagasse, dissolving pulp, wheat straw, rice straw pulp	Bacteria, sugar, biologically treated cellulose, alcohol
Size distribution	Polydispersed	Polydispersed	Polydispersed
Structure	Rod-like/cylindrical particles	Entangled network	3D network
Average size	5–70 nm	5–50 nm	10–100 nm
Process of extraction	Disintegration by hydrolysis of amorphous domains of pure cellulosic substance	Mechanical delamination of softwood pulp in high-pressure homogenizer	Aerobic bacteria
Cross section	Square	Circular	Rectangular
% Crystallinity	54%–88%	51%–69%	84%–89%
Length	100–300 nm	50 nm–3 μm	200 nm–5 μm

adsorption capacities, CNC has a number of other advantages and disadvantages. Some of these include limitations on the operating temperature range, limited wettability, significant moisture absorption, and inconsistency with the majority of polymeric matrix [19]. In the past ten years, there has been a rise in interest in cellulose nanowhisker reinforced composites as a means of overcoming the constraints of CNC. Additionally, progress has been made in controling the thermal and mechanical characteristics of cellulose nanomaterials.

4.3.1.2 Cellulose nanofibers

Typically, CNFs are produced by rigorous mechanical shearing procedures such as high-pressure homogenization, grinding, and refining applied to cellulose pulp fibers. These procedures promote the slow release of constitutive microfibrils, which serve as the fundamental constituents of CNFs. By homogenizing sufficient pressure, Turbak *et al* became the first people to separate CNFs from wood pulp successfully [20]. Several other pre-treatments, including oxidation, acidic, and enzymatic treatment, have been suggested to make the breakdown of the fibers easier (figures 4.4(a) and (b)). These pre-treatments introduce a repulsive charge, weaken the hydrogen bonds, lower DP, and remove amorphous connections between CNFs. The overarching goal of these pre-treatments is to make the hydrogen bonds less stable [21]. The approach that was utilized, as well as the kind of pre-treatment that was performed, directly impacts the dimensions of the CNF. The homogenization of cellulose fibers under high pressure results in the fabrication of CNFs with diameters of 20–40 nm and lengths of several micrometers. When CNF is generated by

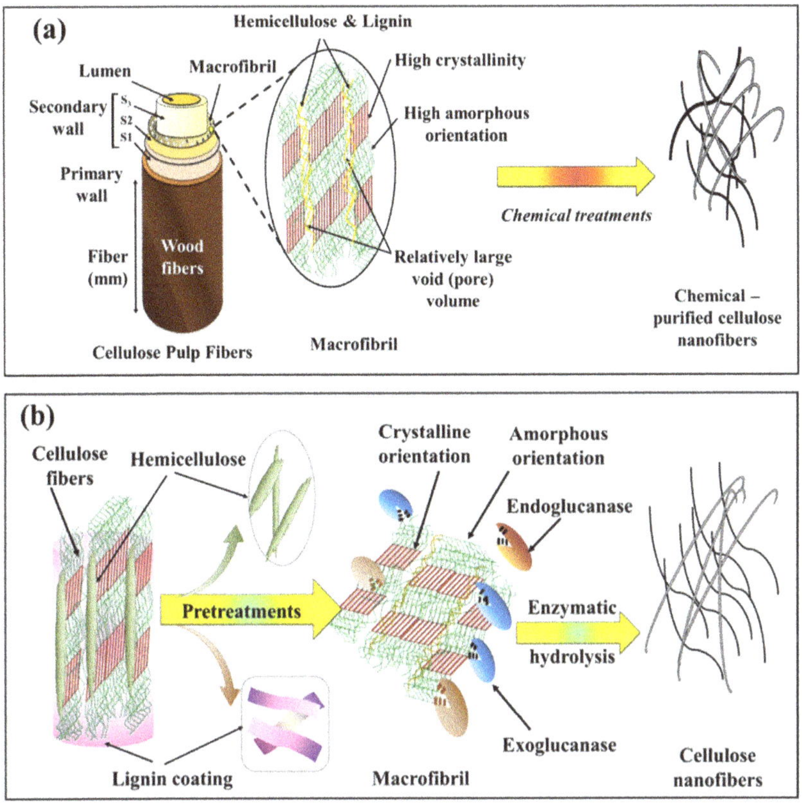

Figure 4.4. (a) Schematic of the steps involved in the chemical process to produce cellulose nanofibers. (b) Schematic representation of the enzymatic pre-treatment involved in the biological process for producing cellulose nanofibers. (Reproduced with permission from [27]. Copyright 2021 Elsevier.)

pre-treating the pulp with an oxidizing agent such as TEMPO (2,2,6,6-tetramethyl-1-piperidinyloxy)-mediated oxidation, the value drops down to 3–5 nm [22]. Before homogenization, the cellulose pulp goes through an acid hydrolysis pre-treatment, giving CNF a finer structure than possible with mechanical treatment. This is because mechanical treatment alone does not include acid hydrolysis [23]. CNFs were isolated from date palm fruit stalk pulp that had been processed with xylanase using a Masuko grinder [24]. The quantity of hemicellulose was reduced as a result of pre-treatment with xylanase, while the polymerization degree of the fibers increased. The findings revealed that using xylanase pre-treatment on the fibers made isolating CNF much simpler. Acid hydrolysis or endoglucanases were applied to wood fiber pulp by Henriksson *et al* along with mechanical disintegration to separate CNF from wood fiber [25]. The disintegration of the fibers was demonstrated to be aided by the enzymatic pre-treatment, and the resulting CNF had a higher average molar mass and larger aspect ratio compared to nanofibers produced just by an acidic pre-treatment [26].

4.3.1.3 Bacterial nanocellulose

Bacterial nanocellulose (BNC), also called bacterial cellulose (BC), is an all-natural polymer of exceptionally high quality. It has a wide variety of possible applications in the area of technology. A J Brown was the first person to describe BNC in a scientific study published in 1886. In order to produce nanocellulose, Brown stated that *Bacterium acetic* produced BNC by fermentative processes [28]. The molecular formula of BNC is identical to that of plant cellulose; however, it is distinguished by a unique three-dimensional (3D) porous network structure. The fact that it is free of lignin and hemicellulose gives it a high degree of purity. Additionally, it is hydrophilic and contains 99% water [29]. It possesses a high polymerization degree (20 000) and a considerable crystallinity (80%), which contribute to its excellent thermal stability. Finally, BNC demonstrates a significant degree of flexibility while having a Young's modulus value for a single nanofiber of 118 GPa [30], which is equivalent to steel. Compared to plant cellulose, BNC nanofibers have a larger surface area [31] and a high aspect ratio, with widths ranging from 20 to 100 nm and lengths ranging from 1 to 9 m. Pellicle, disc, and aggregate are all examples of the numerous forms BNC may take, which contribute to the material's high degree of adaptability. Lastly, because of its exceptionally high purity, which eliminates the need for additional purifying processes, BNC is better for the environment than its vegetable equivalent. In addition, numerous types of waste biomass were recognized as viable carbon sources for the potential industrial fabrication of BNC. This fact, together with the material's high environmental biodegradability and biocompatibility, contribute to its status as an environmentally benign substance. The characteristics of the BNC are affected by the circumstances of fermentation, which include the nitrogen and carbon supplies, the temperature, the length of time spent incubating, and the agitation. In addition, it is feasible to change the structure of the BNC as well as the morphology of the developing fibers by including chemical reagents in the growth media or the fibers themselves [32, 33]. By doing so, it is feasible to generate a diverse template material that may be used for a variery of applications, ranging from medical devices to cutting-edge materials with electrical characteristics [33, 34].

BC provides several benefits, including high crystallinity and purity, high porosity, favorable biocompatibility, durable mechanical qualities, low density, low cost, high absorption capacity, and three-dimensional connected structures with ultrafine nanofibers [35]. Due to its excellent crystallinity and exceptional hydrophilicity, BC has outstanding physical and chemical stabilities and is a promising catalytic support for use in bodies of water [36]. Both of these properties are prsent because BC has a high number of hydroxyl groups [37]. BNC is another member of the family of renewable raw materials and natural biopolymers. It is analogous to nanoscale forms of cellulose, such as CNFs and CNCs, which exhibit enormous promise for environmental and water treatment. This was brought to light in recent research [38].

To summarize this section, nanocellulose is currently the subject of extensive research, and technological advances are being created as a direct consequence of this research. Most research efforts have concentrated on using cellulose nanomaterial as useful candidates for the reinforcements of paper or polymer matrices [39]. On the other hand, it is essential to bring attention to the fact that their production is currently being done on an industrial level and that more current perspective uses may be identified in the relevant published literature. Nanocellulose possesses high mechanical strength, specific surface area, biodegradability, hydrophilicity, and the ability to surface functionalize. These characteristics have increased attention for the potential use of nanocellulose in water treatment.

4.3.2 Chitin/chitosan: chemistry and properties

After cellulose, the most prevalent type of naturally occurring biopolymer is chitin. Chitin turns soluble in acidic environments and is known as chitosan whenever the deacetylation degree reaches around 50%. It is an essential component of the exoskeletons of insects, spiders, and crustaceans and may be found in the cell walls of fungi. The main structural distinction between cellulose and chitin is that each glucose unit's C-2 carbon in chitin bears a-$NHCOCH_3$ group in place of the OH group [40, 41]. The primary source of chitin is the waste from shellfish processing. Commercial uses for chitin waste include the creation of biodegradable plastic food wraps and removing industrial trash. Not only is chitin a rich supply of biomass, as well as a source of specialized biopolymers that can be exploited to develop cutting-edge biomaterials. Chitin provides considerable problems for the modification processes required to develop well-defined derivatives due to being insoluble in common solvents [42].

Chitosan is a natural polysaccharide that is cationic, bioactive, and generated by the attachment of glucosamine and *N*-acetylglucosamine subunits to one another through (1–4) connections. Chitosan is largely characterized using FTIR spectroscopy, which measures the rate of deacetylation [43]. Understanding the possible uses of polymers requires several steps, one of which is identifying the polymer's molecular weight, particularly in biological systems. A number of factors, including molecular weight, deacetylation degree, ionization, and pH, determine the viscosity of chitosan. As the temperature increases, there is a general trend toward reducing the chitosan solution's viscosity [44]. Chitosan can be dissolved in some organic acids (such as formic and acetic acid), but it remains undissolved in organic solvents and water. Chitosan is utilized in numerous practical industrial and biomedical applications owing to its unique physical and chemical characteristics [45, 46]. Figure 4.5 illustrates the process of extracting chitin using naturally occurring resources and converting it into chitosan, along with some of the most important features in its final form. Research conducted in recent years has made it possible to generate chitosan-based nanohybrid—and/or biomaterials that could be used for corrosion prevention, extracting dangerous heavy metals, and antimicrobial packaging [47]. Chitosan, in this context, was combined with a wide variety of different polymers, including polycarbonate, poly(methyl-methacrylate), polystyrene, polysulfide, and polyaniline [48]. Chitosan blends and composites exhibit better characteristics of the material, such as surface area, morphology, mechanical strength, electrical conductivity, bioactivity, and photoluminescence properties.

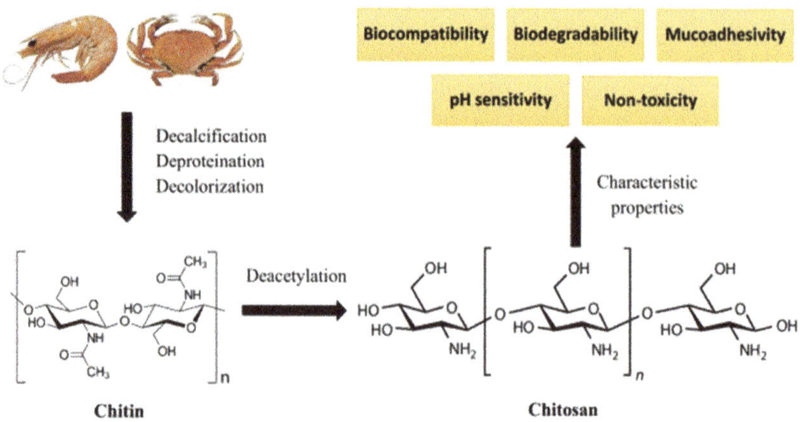

Figure 4.5. Structure of chitin/chitosan and properties. (Reproduced from [50]. Creative Commons Attribution 4.0 International License http://creativecommons.org/licenses/by/4.0/.)

These properties allow their application in diverse fields, including biomedical and antimicrobial materials, biosensors, membrane technology, textiles, and packaging industries [49].

4.3.3 Starch: chemistry and properties

Starch, a commonly used hydrocolloid biopolymer, is an inexpensive polysaccharide that is readily accessible. Granules that are water-loving and hydrophilic are the end product of starch production by agricultural plants. Potatoes, wheat, corn, and rice are the four most common sources of starch. Amylose makes up the majority (80%) of starch, whereas amylopectin accounts for 20%, lipids make up 0.6%, and starch also contains a trace amount of protein and minerals (<0.4%) [51]. Based on the source of the starch, amylopectin and amylose have different relative ratios and molar weights, which leads to the creation of materials with a variety of mechanical properties and levels of biodegradability [52] (figure 4.6). Amylose cannot dissolve in water, while amylopectin can [53]. Amylose cannot be converted into amylopectin. In addition, starch contains a negligible quantity of molecules collectively referred to as 'intermediate compounds'. These molecules have characteristics with both amylose and amylopectin [54], and they make up a very small percentage of the total. The molecular weight of amylose ranges between 10^5 and 10^7 Da, whereas the molecular weight of amylopectin is between 10^7 and 10^9 Da. In addition, the molecular weight of the molecule in the intermediate stage is lower than that of amylopectin but greater than that of amylose [54]. The amount of amylose in the product benefits the elongation and strength of the starch. Enzymes are responsible for the hydrolysis at the acetal link, which is a necessary step in the biodegradation of starch [55]. Amylases are the enzymes that break down the α-1,4 link, and glucosidases are the enzymes that break down the α-1,6 link. The products of deterioration are not harmful in any way. Thermoplastic starch, also known as plasticized starch, is an attractive substitute for synthetic polymers in some

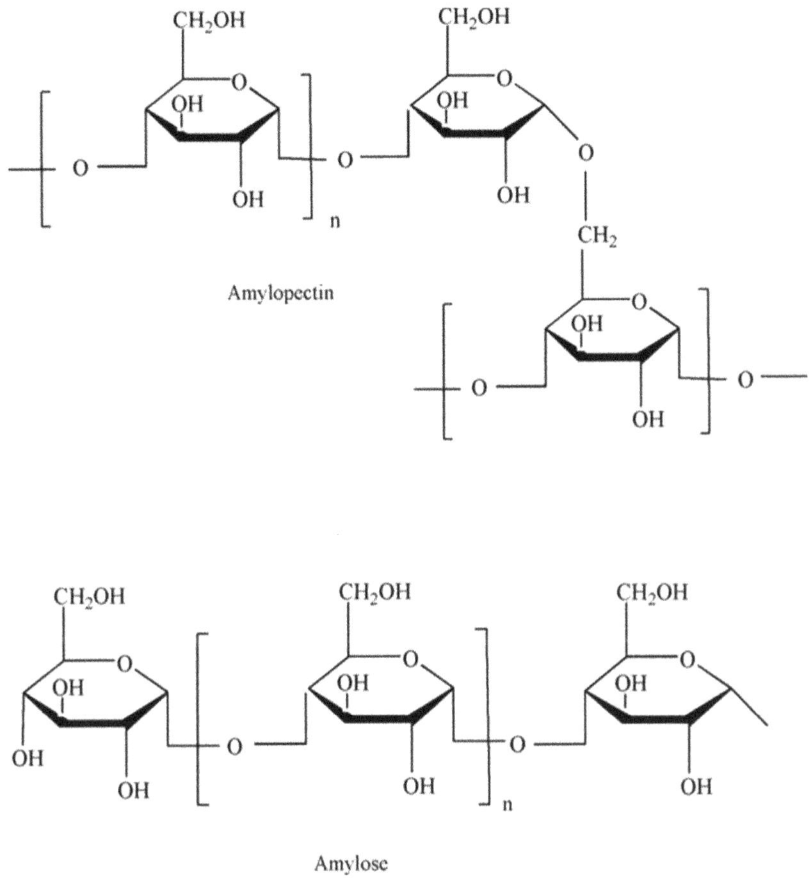

Figure 4.6. Structure of amylopectin and amylose in starch. (Reproduced with permission from [57]. Copyright 2013 Taylor and Francis.)

applications. One example of such an application is the production of starch-based composites. A new category of 'green' composites that are completely biodegradable and go by the name biocomposites is the subject of significant study [56]. They are made of biodegradable polymers reinforced with natural fibers that can also be broken down naturally. Corn, potatoes, and rice are the primary agricultural products utilized in starch manufacturing. Granules are the form in which all of these plants generate starch; nevertheless, the size and content of the granules produced by each plant are distinct from the other plants. In recent years, traditional film-forming resins have seen their prices rise along with their availability, which has led to an increase in the usage of starch as a raw ingredient in the film manufacturing industry. Because starch films have such a low permeability, they make for desirable materials to use in the packaging of food products. Because it deteriorates into innocuous by-products when it comes into contact with the microorganisms that live in the soil, starch can also be effective for producing agricultural mulch films. According to the European bioplastic market statistics, starch might fulfill around

21.3% of the worldwide biopolymer requirement in 2019, which is indicative of the rising demand over the past several years.

4.3.4 Gum: chemistry and properties

Gums are green polymers that are environmentally beneficial, and a variety of goods are made from these natural polysaccharides. Guar gum (GG) is a natural polysaccharide derived from the guar bean's endosperm that is biodegradable, mucoadhesive, and water-soluble. It is made up of D-mannopyranosyl chains linked together by D-galactopyranosyl units. Acacia gum, often known as gum arabic (GA), is a complex exudate produced by the *Acacia senegal* and *Acacia seyal* trees. It is a branched heteropolysaccharide. One of the earliest gums known to science, GA is made up of D-galactopyranosyl units. GA has various applications in the food and pharmaceutical industries [35]. Because of their excellent capacity to modify rheological characteristics, GA and GG have found widespread application as ion exchange resins, binders, thickening agents, and dispersion agents [58]. In addition to serving as an emulsifying, additive, reducing, and potential stabilizer in the preparation of various catalysts, they have also been used as biosorbents in a variety of applications and for reducing or stabilizing nanomaterials. They have also been designed to stabilize magnetic nanoparticles effectively. Different applications of GG are depicted in figure 4.7.

Pseudomonas elodea produces gellan gum, a polysaccharide soluble in water and has a negative charge. In most cases, it is composed of a tetra-saccharide, which comprises two D-glucose residues and one residue of L-rhamnose and D-glucuronic acid. On the market, gelan gum is one of the hydrocolloids with the highest price [60]. Gellan gum is a material that may be inserted into tissue and does not have any harmful effects. It has been utilized in humans *in vivo* as a vehicle for the administration of eye medicines. Gellan is a popular choice for use as a gelatinous agent. It may be employed to generate patterned solutions, which are highly powerful suspended agents. The gelling systems that make up these patterned solutions are sheared either during or after the formation of the gel [61]. Compared to other polysaccharide hydrogels such as agar, alginate, and carrageenan, gellan gum hydrogel has a reduced dependence on pH. As a consequence, the usage of gellan gum in the food and pharmaceutical industries is quite extensive. Additionally, gellan gum was researched for its possible use in the delivery of antimicrobial medications to wounds and as a nontoxic tissue engineering scaffold. As a result, gellan gum can be used as a flexible dressing that can be applied as a hydrogel sheet or as an *in situ* liquid that promotes gelling. There is a shortage of published data on the effectiveness of gellan gum *in vivo*, although it is an appealing raw material for the healing process [62, 63].

A naturally occurring polysaccharide called xanthan gum is produced by the bacterium *Xanthomonas campestris*. The fundamental structure of xanthan is comprised of repeated pentasaccharide units, each composed of one D-glucopyranosyluronic unit, two D-mannopyranosyl units, and two D-glucopyranosyl units [64]. The degree to which xanthan solutions can thicken the proportional to their

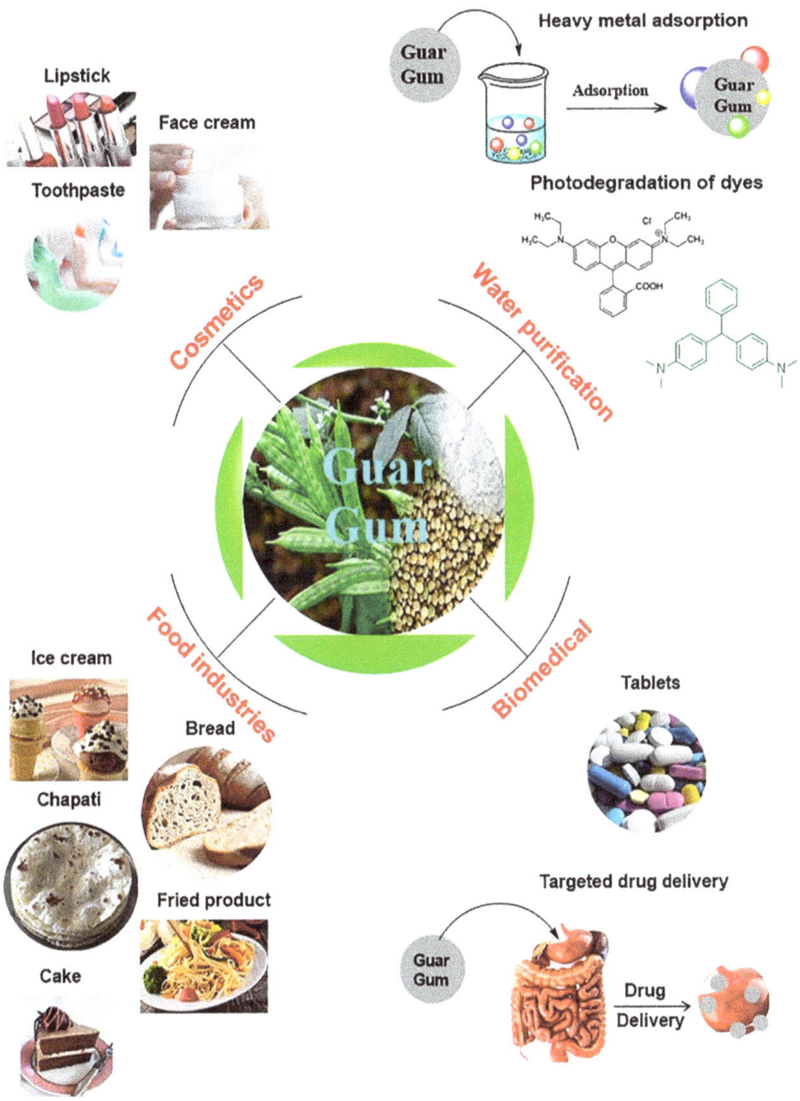

Figure 4.7. Various applications of guar gum. (Reproduced with permission from [59]. Copyright 2018 Elsevier.)

viscosity; solutions with high viscosity are more resistant to flow. Xanthan solutions are pseudoplastic and their viscosity decreases with shear rate.

Xanthan gum is a common ingredient in both toothpaste and cosmetics as it is not difficult to squeeze out of a tube. In addition, it guarantees that the toothpaste will maintain its firm standing position on the brush. Because of the shear thinning qualities, the dispersion on and washing off of the teeth are both improved. The xanthan gum used to thicken toothpaste produces a brilliant, glossy cord and has limited flow behavior. When used in pharmaceutical suspensions or emulsions, xanthan gum helps to avoid the separation of insoluble components [65].

4.3.5 Alginate: chemistry and properties

The anionic polysaccharide known as alginate, also known as algin or alginic acid, has linear chains of α-L-guluronic acid (G) and 1,4′-linked β-D-mannuronic acid (M) (figure 4.8). Brown algae have large amounts of alginate scattered throughout their cell walls, where it binds water to produce a sticky glue. In its extracted state, it can absorb 200–300 times its weight in water and has a high rate of water absorption. Alginic acid is available in powdered, granular, and filamentous forms and can range in color from white to yellowish-brown. Its chemical name is 'alginate'. It is a biocompatible polymer that is inexpensive, nontoxic, and biodegradable. It also demonstrates several unique qualities, such as the absorption of metal ions, capacity to exchange ions, and promotion of wound healing. Therefore, this natural product has various uses in the pharmaceutical and food sectors, such as in the fabrication of artificial skin, surgical tape, and tissue engineering material [66]. Because of its high usefulness, alginic acid is utilized in a variety of applications within the biomedical industry. When divalent cations such as calcium ions are present, alginic acid will spontaneously undergo a minor gelling process. This is because alginic acid has a high percentage of acid. This gentle gelling characteristic enables the encapsulation of a wide variety of compounds and even cells within alginate gels with little detrimental effect [67]. The rate of drug release can be changed by altering the

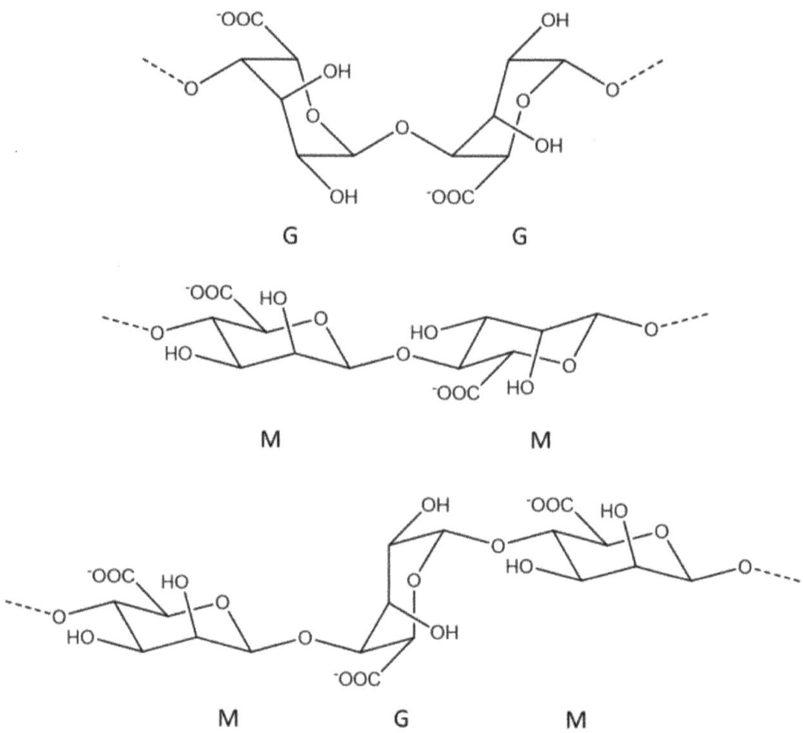

Figure 4.8. Chemical structures of G-block, M-block, and alternating block in alginate. (Reproduced with permission from [69]. Copyright 2012 Elsevier.)

interaction between the polymer and the drug as well as by chemically immobilizing the drug to the polymer's backbone by using reactive carboxylate groups, according to extensive research on the use of alginate as a drug delivery device. Encapsulating bioactive and protein substances in ionically cross-linked alginate gels improves their efficacy and targetability, hence alginate gels have been studied extensively to produce protein delivery systems [68].

4.3.6 Pectin: chemistry and properties

Land plants have pectin, a type of structural heteropolysaccharide, in their primary cell walls. Pectin comprises an intricate group of polysaccharides found in most main cell walls. These polysaccharides are found in especially high concentrations in the non-woody sections of almost all terrestrial plant species. Pectin used in commercial applications is often made from apple pomace or citrus peel, both of which are waste products generated during the production of fruit juices. Pectin is assumed to be made up mostly of D-galacturonic acid (GalA) units, which are linked together in chains by an α-(1–4) glycosidic bond (figure 4.9) [70]. In the process of manufacturing mucoadhesive patches, pectin is typically combined with carbopol and chitosan. It exerts its effect by making close contact with the tissues, which causes the tangling of polymer and mucin chains and the development of weak connections between the chains. Pectin is a natural chemical that may be used as a prophylactic agent to protect against poisoning with poisonous cations. Pectin is an excellent agent for eliminating mercury and lead from the gastrointestinal system as well as the respiratory organs. Pectin helps the blood to coagulate better, and therefore is helpful in reducing bleeding [70]. Pectin is an effective therapy for overeating because it slows digestion by immobilizing food components in the colon, which leads to less food being absorbed [70]. Pectin hydrogels have been utilized in the process of tablet formulation both as a binding agent and in the controlled release matrix tablet formulation process. Pectin is now being investigated for potential use in colon-targeted medication delivery systems [71].

There was some evidence that some pectins might reduce inflammation. *In vitro*, this is a desirable and encouraging quality in the tuning of medical devices. Pectins are now the subject of extensive research in biomaterials as potential new options for

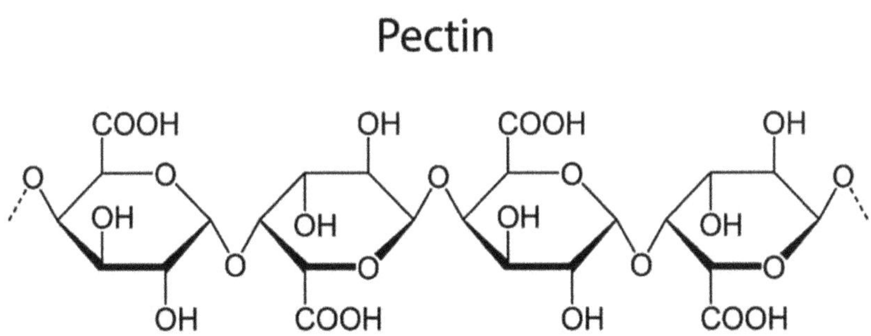

Figure 4.9. The structure of pectin.

hard and soft tissue engineering as well as dental titanium coating due to the fact that they stimulate *in vitro* immune responses [71].

4.4 Characterization of green polymers

4.4.1 UV–vis spectroscopy

The use of UV–vis spectroscopy is critical for determining the identity of nanomaterials as well as analysing and characterizing them. In order to characterize different biopolymeric nanomaterials with sizes ranging from 2 to 100 nm, light with a wavelength range of 300–800 nm is frequently utilized [72]. The shape, size, and particle interaction with the medium all influence the shift in the surface plasmon resonance (SPR) bands.

4.4.2 X-ray diffraction (XRD)

The amorphous and crystalline nature of biopolymeric materials can be determined using the XRD technique. It details the methodology that will be used to quantify cementation materials. XRD analysis can be used to find out the way atoms are arranged, the size of crystallites, and any imperfections in crystalline materials. For the XRD test, a source of Cu Kα radiation (11/4 1.5406 Å) is used. The Bragg–Brentano method is used to evaluate and name crystalline substances that are not well known [73, 74]. This technique has the potential to determine the components and phases of a solid since every solid has a unique x-ray diffraction pattern. Using specialized detectors, the XRD method measures the intensity of the rays that have been bent after being focussed at an initial angle of 2θ on a biopolymer film or membrane. After the measurement, the incident ray's angle is altered by 2θ for a new reading leading to a final value of 2θ. Bragg's law governs the crystalline phases' diffracted beam. Biopolymer film testing often uses a copper radiation source (Cu Kα) with a 2 scan range from 5° to 70° (0.20° min^{-1}–0.30° min^{-1} scanning speed) [74].

4.4.3 X-ray photoelectron spectroscopy (XPS)

XPS, also known as electron spectroscopy for chemical analysis (ESCA), is a very sensitive quantitative spectroscopy method. This method is utilized to analyze the surface chemistry, chemical state, and elemental composition of nanomaterials.

4.4.4 Raman spectroscopy

In recent years, the Raman spectroscopy approach has been one of the most common vibrational spectroscopic techniques utilized to provide information about crystallinity, chemical structure, phase, and molecular interactions. In the Raman spectroscopy technique, a laser light source is used to irradiate a sample of material, generating an infinitesimal quantity of Raman scattered light. A CCD camera detects this light as a Raman spectrum [75].

4.4.5 Scanning electron microscope (SEM)

Instead of using light to illuminate images, SEM uses electrons. An electron beam interacting with an atom sample produces a variety of signals, which are then gathered by detectors, and ultimately result in the creation of a sample image. In order to account for the non-conductive character of biopolymers in SEM investigations, a tiny layer of metal must be sputtered on their surface [76]. The SEM method is used for fracture and surface analysis in biopolymer membranes and films. This approach enables the assessment of sample morphological and structural aspects. When biopolymer membranes and films are formed using the casting method [77], the surface of the membrane or film is typically consistent and smooth throughout its whole. Under some conditions, the SEM can reveal a porous structure on the surface of the material if the film was prepared using a certain method and/or the biopolymer mixture was carried out. According to recent research, SEM may be useful for validating the material structure of multilayer films made of silk fibroin and gelatin. This technique is significant because it provides three-dimensional images of the interior of the eye, which helps researchers gain a better understanding of the surface morphology of the eye's component parts [78, 79].

4.4.6 Energy dispersive x-ray spectroscopy (EDX)

The elemental analysis method is known as EDX. It is connected to electron microscopy and is based on the production of distinctive x-rays that reveal the chemical composition and presence of elements in a sample of nanomaterials [80].

4.4.7 Transmission electron microscopy (TEM)

TEM images are generated due to electrons interacting with various sources after being passed through a testing material sample employing a beam of electrons. Because of the three-dimensional effect, TEM images provide views of the surface that are richer in detail. Compared to photographs taken with an SEM, pictures obtained by TEM analysis offer a more magnified view of the material. The magnification and three-dimensional analytical capabilities of TEM make it possible to understand the surface morphology and the behavior of the contacts before and after the interactions. Using TEM images, one may also assess the thickness of the material being tested on its outside surface [81, 82]. The TEM method can perform a microstructural investigation of biopolymer membranes and films that are not detectable by the SEM method. This includes the investigation of microporous structures that are generated in a modified film production process. Investigating nonreinforcement dispersion in biopolymer composites is another use for TEM [83].

4.4.8 Fourier transform infrared spectroscopy (FTIR)

Infrared spectrometry is employed to investigate how different types of radiation, particularly infrared radiation in the electromagnetic spectrum, interact with different types of materials. Infrared spectroscopy is a method that may be utilized to distinguish and characterize various chemical compounds and materials. Fourier transform infrared spectroscopy (FTIR) may be applied in biopolymers to explore the various functional groups and interactions. When investigating the spectra [84], one may easily identify hydrogen bonds, amide links, and other types of interactions. The attenuated total reflection (ATR) sampling device-based FTIR spectroscopy method is one of the research approaches currently being investigated. It requires little sample preparation, enables simultaneous estimates in the laboratory as well as in the field, and is straightforward to employ. The thermotropic phase behavior of a series of recently synthesized self-forming synthetic biopolymers [85] was investigated using variable-temperature FTIR absorption spectroscopy. Numerous studies have used the FTIR technique to examine the various mechanical and physical properties of biopolymers. Two-dimensional correlation mapping was used to examine the ATR spectral intensity variations of immature and mature cotton fibers [86]. Measurements taken with an FTIR spectrometer may also be used to confirm the interactions between the various biopolymers that make up its composite ingredients [87]. To identify the specific functional group of the biopolymers, the vibrational ranges of the biopolymers were investigated, including the absorption of ketones, R–CO–R, at a wavelength of 1730–1740 cm^{-1} in the case of the biopolymers. R–COOH carboxylic acids may be identified by their two unique bands, which can be seen around 3500 cm^{-1} and 1700 cm^{-1}, respectively. These bands correspond to the O–H and C = O stretching vibrations of the carboxyl group (–COOH). The absorption of metal carboxylates or carboxylic salts occurs between 1610 cm^{-1} and 1550 cm^{-1}, whereas the absorption of saturated or aromatic carboxylic acids occurs between 1680 and 1690 cm^{-1} [88].

References

[1] Peramune D, Manatunga D C, Dassanayake R S, Premalal V, Liyanage R N, Gunathilake C and Abidi N 2022 Recent advances in biopolymer-based advanced oxidation processes for dye removal applications: a review *Environ. Res.* **215** 114242

[2] Suresh R, Rajendran S, Hoang T K A, Vo D V N, Siddiqui M N and Cornejo-Ponce L 2021 Recent progress in green and biopolymer based photocatalysts for the abatement of aquatic pollutants *Environ. Res.* **199** 111324

[3] Banerjee A and Bandopadhyay R 2016 Use of dextran nanoparticle: a paradigm shift in bacterial exopolysaccharide based biomedical applications *Int. J. Biol. Macromol.* **87** 295–301

[4] Farshchi E, Pirsa S, Roufegarinejad L, Alizadeh M and Rezazad M 2019 Photocatalytic/biodegradable film based on carboxymethyl cellulose, modified by gelatin and TiO_2–Ag nanoparticles *Carbohydr. Polym.* **216** 189–96

[5] Kolangare I M, Isloor A M, Karim Z A, Kulal A, Ismail A F, Inamuddin and Asiri A M 2019 Antibiofouling hollow-fiber membranes for dye rejection by embedding chitosan and silver-loaded chitosan nanoparticles *Environ. Chem. Lett.* **17** 581–7

[6] Crini G, Torri G, Lichtfouse E, Kyzas G Z, Wilson L D and Morin-Crini N 2019 Dye removal by biosorption using cross-linked chitosan-based hydrogels *Environ. Chem. Lett.* **17** 1645–66

[7] Lichtfouse E *et al* 2019 Chitosan for direct bioflocculation of wastewater *Environ. Chem. Lett.* **17** 1603–21

[8] Vanaamudan A, Sadhu M and Pamidimukkala P 2018 Chitosan–guar gum blend silver nanoparticle bionanocomposite with potential for catalytic degradation of dyes and catalytic reduction of nitrophenol *J. Mol. Liq.* **271** 202–8
[9] Morin-Crini N, Lichtfouse E, Torri G and Crini G 2019 Applications of chitosan in food, pharmaceuticals, medicine, cosmetics, agriculture, textiles, pulp and paper, biotechnology, and environmental chemistry *Environ. Chem. Lett.* **17** 1667–92
[10] Raveendran P, Fu J and Wallen S L 2003 Completely 'green' synthesis and stabilization of metal nanoparticles *J. Am. Chem. Soc.* **125** 13940–1
[11] Shankar S and Rhim J W 2020 Bionanocomposite films for food packaging applications *Innovative Food Processing Technologies: A Comprehensive Review* (Amsterdam: Elsevier) pp 234–43
[12] Sarkar S, Ponce N T, Banerjee A, Bandopadhyay R, Rajendran S and Lichtfouse E 2020 Green polymeric nanomaterials for the photocatalytic degradation of dyes: a review *Environ. Chem. Lett.* **18** 1569–80
[13] Olivera S, Muralidhara H B, Venkatesh K, Guna V K, Gopalakrishna K and Kumar Y K 2016 Potential applications of cellulose and chitosan nanoparticles/composites in wastewater treatment: a review *Carbohydr. Polym.* **153** 600–18
[14] Postek M, Moon R J, Rudie A W and Bilodeau M A 2013 *Production and Applications of Cellulose Nanomaterials* (Peachtree Corners, GA: TAPPI)
[15] Abouzeid R E, Khiari R, El-Wakil N and Dufresne A 2019 Current state and new trends in the use of cellulose nanomaterials for wastewater treatment *Biomacromolecules* **20** 573–97
[16] Kim J H, Shim B S, Kim H S, Lee Y J, Min S K, Jang D, Abas Z and Kim J 2015 Review of nanocellulose for sustainable future materials *Int. J. Precis. Eng. Manuf.-Green Technol.* **2** 197–213
[17] Habibi Y, Lucia L A and Rojas O J 2010 Cellulose nanocrystals: chemistry, self-assembly, and applications *Chem. Rev.* **110** 3479–500
[18] Mohammed N 2017 Cellulose nanocrystals incorporated nanocomposites for water treatment applications *Doctoral Thesis* University of Waterloo, ON https://uwspace.uwaterloo.ca/handle/10012/11168
[19] Siaueira G, Bras J and Dufresne A 2009 Cellulose whiskers versus microfibrils: influence of the nature of the nanoparticle and its surface functionalization on the thermal and mechanical properties of nanocomposites *Biomacromolecules* **10** 425–32
[20] Turbak A F, Snyder F W and Sandberg K R 1983 Microfibrillated cellulose, a new cellulose product: properties, uses, and commercial potential *J. Appl. Polym. Sci., Appl. Polym. Symp.* **37** 815–27
[21] Lavoine N, Desloges I, Dufresne A and Bras J 2012 Microfibrillated cellulose—its barrier properties and applications in cellulosic materials: a review *Carbohydr. Polym.* **90** 735–64
[22] Eichhorn S J *et al* 2010 Review: current international research into cellulose nanofibres and nanocomposites *J. Mater. Sci.* **45** 1–33
[23] Zimmermann T, Bordeanu N and Strub E 2010 Properties of nanofibrillated cellulose from different raw materials and its reinforcement potential *Carbohydr. Polym.* **79** 1086–93
[24] Hassan M L, Bras J, Hassan E A, Silard C and Mauret E 2014 Enzyme-assisted isolation of microfibrillated cellulose from date palm fruit stalks *Ind. Crops Prod.* **55** 102–8
[25] Henriksson M, Henriksson G, Berglund L A and Lindström T 2007 An environmentally friendly method for enzyme-assisted preparation of microfibrillated cellulose (MFC) nanofibers *Eur. Polym. J.* **43** 3434–41

[26] Tshikovhi A, Mishra S B and Mishra A K 2020 Nanocellulose-based composites for the removal of contaminants from wastewater *Int. J. Biol. Macromol.* **152** 616–32
[27] Sayyed A J, Pinjari D V, Sonawane S H, Bhanvase B A, Sheikh J and Sillanpää M 2021 Cellulose-based nanomaterials for water and wastewater treatments: a review *J. Environ. Chem. Eng.* **9** 106626
[28] Brown A J 1886 XLIII.—On an acetic ferment which forms cellulose *J. Chem. Soc. Trans.* **49** 432–9
[29] Qiu K and Netravali A N 2014 A review of fabrication and applications of bacterial cellulose based nanocomposites *Polym. Rev.* **54** 598–626
[30] Soriano M L and Dueñas-Mas M J 2018 Promising sensing platforms based on nanocellulose *Carbon-Based Nanosensor Technology* (Berlin: Springer) pp 273–301
[31] Foresti M L, Vázquez A and Boury B 2017 Applications of bacterial cellulose as precursor of carbon and composites with metal oxide, metal sulfide and metal nanoparticles: a review of recent advances *Carbohydr. Polym.* **157** 447–67
[32] Hu W, Chen S, Yang J, Li Z and Wang H 2014 Functionalized bacterial cellulose derivatives and nanocomposites *Carbohydr. Polym.* **101** 1043–60
[33] Torres F G, Arroyo J J and Troncoso O P 2019 Bacterial cellulose nanocomposites: an all-nano type of material *Mater. Sci. Eng.* C **98** 1277–93
[34] Andriani D, Apriyana A Y and Karina M 2020 The optimization of bacterial cellulose production and its applications: a review *Cellulose* **27** 6747–66
[35] Nasrollahzadeh M, Sajjadi M, Iravani S and Varma R S 2021 Starch, cellulose, pectin, gum, alginate, chitin and chitosan derived (nano)materials for sustainable water treatment: a review *Carbohydr. Polym.* **251** 116986
[36] Fang Q, Zhou X, Deng W, Zheng Z and Liu Z 2016 Freestanding bacterial cellulose–graphene oxide composite membranes with high mechanical strength for selective ion permeation *Sci. Rep.* **6** 33185
[37] Costa S V, Gonïalves A S, Zaguete M A, Mazon T and Nogueira A F 2013 ZnO nanostructures directly grown on paper and bacterial cellulose substrates without any surface modification layer *Chem. Commun.* **49** 8096–8
[38] Wang D 2019 A critical review of cellulose-based nanomaterials for water purification in industrial processes *Cellulose* **26** 687–701
[39] Dufresne A 2012 Nanocellulose: from nature to high performance tailored materials *Holzforschung* **67** 353
[40] Dash M, Chiellini F, Ottenbrite R M and Chiellini E 2011 Chitosan—a versatile semi-synthetic polymer in biomedical applications *Prog. Polym. Sci.* **36** 981–1014
[41] Sánchez R, Stringari G B, Franco J M, Valencia C and Gallegos C 2011 Use of chitin, chitosan and acylated derivatives as thickener agents of vegetable oils for bio-lubricant applications *Carbohydr. Polym.* **85** 705–14
[42] Kurita K, Kojima T, Nishiyama Y and Shimojoh M 2000 Synthesis and some properties of nonnatural amino polysaccharides: branched chitin and chitosan *Macromolecules* **33** 4711–6
[43] Sivashankari P R and Prabaharan M 2017 Deacetylation modification techniques of chitin and chitosan *Chitosan Based Biomaterials* vol 1 (Amsterdam: Elsevier) pp 117–33
[44] Cho J, Heuzey M C, Bégin A and Carreau P J 2006 Viscoelastic properties of chitosan solutions: effect of concentration and ionic strength *J. Food Eng.* **74** 500–15

[45] Bilal M, Asgher M, Iqbal M, Hu H and Zhang X 2016 Chitosan beads immobilized manganese peroxidase catalytic potential for detoxification and decolorization of textile effluent *Int. J. Biol. Macromol.* **89** 181–9
[46] Bilal M, Iqbal H M N, Hu H, Wang W and Zhang X 2017 Enhanced bio-catalytic performance and dye degradation potential of chitosan-encapsulated horseradish peroxidase in a packed bed reactor system *Sci. Total Environ.* **575** 1352–60
[47] Bilal M, Zhao Y, Rasheed T, Ahmed I, Hassan S T S, Nawaz M Z and Iqbal H M N 2019 Biogenic nanoparticle–chitosan conjugates with antimicrobial, antibiofilm, and anticancer potentialities: development and characterization *Int. J. Environ. Res. Public Health* **16** 598
[48] Winie T and Mohd Shahril N S 2015 Conductivity enhancement by controlled percolation of inorganic salt in multiphase hexanoyl chitosan/polystyrene polymer blends *Front. Mater. Sci.* **9** 132–40
[49] Kausar A 2019 Polymer and modified chitosan-based nanocomposite: impending material for technical application *Polym. Technol. Mater.* **58** 934–47
[50] Qamar S A, Ashiq M, Jahangeer M, Riasat A and Bilal M 2020 Chitosan-based hybrid materials as adsorbents for textile dyes—a review *Case Stud. Chem. Environ. Eng.* **2** 100021
[51] Alcázar-Alay S C and Meireles M A A 2015 Physicochemical properties, modifications and applications of starches from different botanical sources *Food Sci. Technol.* **35** 215–36
[52] Ratnayake W S, Hoover R, Shahidi F, Perera C and Jane J 2001 Composition, molecular structure, and physicochemical properties of starches from four field pea (*Pisum sativum* L.) cultivars *Food Chem.* **74** 189–202
[53] Bertolini A (ed) 2009 *Starches: Characterization, Properties, and Applications* (London: Routledge)
[54] Vilaplana F, Hasjim J and Gilbert R G 2012 Amylose content in starches: toward optimal definition and validating experimental methods *Carbohydr. Polym.* **88** 103–11
[55] Subach D J 1997 Biodegradable polymers *Chemist* **74** 7–9
[56] Netravali A N and Chabba S 2003 Composites get greener *Mater. Today* **6** 22–9
[57] Ismail H, Irani M and Ahmad Z 2013 Starch-based hydrogels: present status and applications *Int. J. Polym. Mater. Polym. Biomater.* **62** 411–20
[58] Padil V V T, Wacławek S, Černík M and Varma R S 2018 Tree gum-based renewable materials: sustainable applications in nanotechnology, biomedical and environmental fields *Biotechnol. Adv.* **36** 1984–2016
[59] Sharma G, Sharma S, Kumar A, Al-Muhtaseb A H, Naushad M, Ghfar A A, Mola G T and Stadler F J 2018 Guar gum and its composites as potential materials for diverse applications: a review *Carbohydr. Polym.* **199** 534–45
[60] Morales M E and Ruiz M A 2016 Microencapsulation of probiotic cells: applications in nutraceutic and food industry *Nutraceuticals* (Amsterdam: Elsevier) pp 627–68
[61] Fallourd M J and Viscione L 2009 Ingredient selection for stabilisation and texture optimisation of functional beverages and the inclusion of dietary fibre *Functional and Speciality Beverage Technology* (Amsterdam: Elsevier) pp 3–38
[62] Smith A M, Moxon S and Morris G A 2016 Biopolymers as wound healing materials *Wound Healing Biomaterials* (Amsterdam: Elsevier) pp 261–87
[63] Cencetti C, Bellini D, Pavesio A, Senigaglia D, Passariello C, Virga A and Matricardi P 2012 Preparation and characterization of antimicrobial wound dressings based on silver, gellan, PVA and borax *Carbohydr. Polym.* **90** 1362–70

[64] Wang N L, Kiyohara H, Matsumoto T, Otsuka H, Hirano M and Yamada H 1994 Polyclonal antibody against a complement-activating pectin from the roots of *Angelica acutiloba Planta Med.* **60** 425–9
[65] Michaelsen T E, Gilje A, Samuelsen A B, Høgåsen K and Paulsen B S 2000 Interaction between human complement and a pectin type polysaccharide fraction, PMII, from the leaves of *Plantago major* L. *Scand. J. Immunol.* **52** 483–90
[66] Yamada M and Honma I 2004 Alginic acid–imidazole composite material as anhydrous proton conducting membrane *Polymer* **45** 8349–54
[67] Klöck G, Pfeffermann A, Ryser C, Gröhn P, Kuttler B, Hahn H J and Zimmermann U 1997 Biocompatibility of mannuronic acid-rich alginates *Biomaterials* **18** 707–13
[68] Augst A D, Kong H J and Mooney D J 2006 Alginate hydrogels as biomaterials *Macromol. Biosci.* **6** 623–33
[69] Lee K Y and Mooney D J 2012 Alginate: properties and biomedical applications *Prog. Polym. Sci.* **37** 106–26
[70] Chen C H, Sheu M T, Chen T F, Wang Y C, Hou W C, Liu D Z, Chung T C and Liang Y C 2006 Suppression of endotoxin-induced proinflammatory responses by citrus pectin through blocking LPS signaling pathways *Biochem. Pharmacol.* **72** 1001–9
[71] Salman H, Bergman M, Djaldetti M, Orlin J and Bessler H 2008 Citrus pectin affects cytokine production by human peripheral blood mononuclear cells *Biomed. Pharmacother.* **62** 579–82
[72] Sadeghi B, Mohammadzadeh M and Babakhani B 2015 Green synthesis of gold nanoparticles using *Stevia rebaudiana* leaf extracts: characterization and their stability *J. Photochem. Photobiol.* B **148** 101–6
[73] Vishwakarma V and Uthaman S 2019 Environmental impact of sustainable green concrete *Smart Nanoconcretes and Cement-Based Materials: Properties, Modelling and Applications* (Amsterdam: Elsevier) pp 241–55
[74] Das A, Ringu T, Ghosh S and Pramanik N 2022 A comprehensive review on recent advances in preparation, physicochemical characterization, and bioengineering applications of biopolymers *Polym. Bull.* 1–66
[75] Kuhar N, Sil S, Verma T and Umapathy S 2018 Challenges in application of Raman spectroscopy to biology and materials *RSC Adv.* **8** 25888–908
[76] Maus A 2008 Characterization and analysis of polymers *Macromol. Chem. Phys.* **209** 1515
[77] Das Murtey M and Ramasamy P 2016 Sample preparations for scanning electron microscopy—life sciences *Modern Electron Microscopy in Physical and Life Sciences* (London: InTech)
[78] Venkatesan J, Lee J Y, Kang D S, Anil S, Kim S K, Shim M S and Kim D G 2017 Antimicrobial and anticancer activities of porous chitosan–alginate biosynthesized silver nanoparticles *Int. J. Biol. Macromol.* **98** 515–25
[79] Mandal B B, Mann J K and Kundu S C 2009 Silk fibroin/gelatin multilayered films as a model system for controlled drug release *Eur. J. Pharm. Sci.* **37** 160–71
[80] Scimeca M, Bischetti S, Lamsira H K, Bonfiglio R and Bonanno E 2018 Energy dispersive x-ray (EDX) microanalysis: a powerful tool in biomedical research and diagnosis *Eur. J. Histochem.* **62** 89–99
[81] Ghosh S and Ray A 2015 Spontaneous vesicle based excipient formation in mixtures of sodium *N*-(*n*-alkanoyl)-L-alaninate and *N*-cetylpyridinium chloride: effect of hydrocarbon chain length *Ind. Eng. Chem. Res.* **54** 1953–61

[82] Ghosh S, Ray A, Pramanik N and Ambade B 2016 Can a catanionic surfactant mixture act as a drug delivery vehicle? *C. R. Chim.* **19** 951–4
[83] Noishiki Y, Nishiyama Y, Wada M, Kuga S and Magoshi J 2002 Mechanical properties of silk fibroin-microcrystalline cellulose composite films *J. Appl. Polym. Sci.* **86** 3425–9
[84] Theophile T 2015 *Infrared Spectroscopy—Anharmonicity of Biomolecules, Crosslinking of Biopolymers, Food Quality and Medical Applications* (London: InTech)
[85] Bista R K, Bruch R F and Covington A M 2011 Infrared spectroscopic study of thermotropic phase behavior of newly developed synthetic biopolymers *Spectrochim. Acta* A **81** 583–9
[86] Liu Y, Thibodeaux D and Gamble G 2012 Characterization of attenuated total reflection infrared spectral intensity variations of immature and mature cotton fibers by two-dimensional correlation analysis *Appl. Spectrosc.* **66** 198–207
[87] Teixeira E D M, De Campos A, Marconcini J M, Bondancia T J, Wood D, Klamczynski A, Mattoso L H C and Glenn G M 2014 Starch/fiber/poly(lactic acid) foam and compressed foam composites *RSC Adv.* **4** 6616–23
[88] Riaz U and Ashraf S M 2015 Characterization of polymer blends with FTIR spectroscopy *Characterization of Polymer Blends: Miscibility, Morphology and Interfaces* (Oxford: Wiley Blackwell) pp 625–78

IOP Publishing

Photocatalytic Dye Degradation Using Green Polymeric-based Nanostructures

Principles and applications in wastewater treatment

Muhammad Ikram, Muhammad Ahsaan Bari and Junaid Haider

Chapter 5

Photocatalytic dye degradations with green polymeric-based nanomaterials

Green photocatalysts have become a well-known wastewater treatment technology in recent years. Nanostructured materials are combined with biopolymers for improved biodegradability, recycling, and an environmentally friendly approach in order to make future advancements. In this respect, polysaccharide-based metal, metal oxide, and metal sulfide materials are believed to be increasingly important in improving the effectiveness of photocatalysts. They not only enhance the capability for degradation, they also minimize both the toxicity and the cost. In the current chapter, a variety of nanomaterials that have been doped with polysaccharides are investigated and discussed in detail. This chapter focuses mostly on metal, metal oxide, and metal sulfide photocatalysts that are based on chitosan, chitin, cellulose, pectin, gum, and alginate. These photocatalysts have attracted a lot of attention from researchers. In addition, the problems that the world is currently experiencing with the application of nanomaterials for dye degradation are discussed.

5.1 Green photocatalysts

Biopolymers are considered the best option for the removal of dye because of their high levels of affordability, renewability, biodegradability, non-toxicity, and surface tunability. Additionally, biopolymer-based materials for photocatalysis (green photocatalysts), such as protein- and polysaccharide-based metal oxides, have been investigated owing to the synergistic impacts of the chemical and physical connections between the organic and inorganic components. Because of the one-of-a-kind surface properties which result from very small effects, biopolymers and their derivatives have recently seen widespread application on the nanoscale. These properties include increased physicochemical stability and surface area, and a porous structure resembling an extracellular matrix having high meso- and

doi:10.1088/978-0-7503-5517-9ch5

microporosity. The nanobiopolymers that have been utilized effectively in numerous industries, such as the energy sector, catalysis, the treatment of water, the food industry, and the biomedical industry, have been synthesized with various ultra-high aspect and length–diameter ratios [1]. The part that biopolymers play in the photocatalytic process has been under much scrutiny over the past few years. For the process of photocatalysis, the use of biopolymers such as chitin, cellulose, and chitosan as solid substrates has become increasingly common. The advantages and disadvantages of these polymers are summarized in table 5.1. Because of their relatively high availability, flexibility, synergistic effects, and ease of surface functionalization with varying degrees of crystallinity, biopolymers provide significant advantages when used as a substrate. Biopolymers can act as potent stabilizing and reducing agents owing to the hydroxyl groups on their surfaces [2]. In addition to this, biopolymers have the ability to inhibit the self-aggregation of nanomaterials, hence keeping their morphology and shape. In this chapter we present some current examples of biopolymer-based substrates, including chitin, cellulose, and chitosan,

Table 5.1. Advantages and disadvantages of green polymers in photocatalytic dye removal techniques [1].

Green polymer	Advantages	Disadvantages
Chitin/ chitosan and cellulose	Prevent photocatalyst loss. High surface area and porosity for catalyst active sites. Increase recyclability. Inexpensive. Excellent chelating agents. Act as carriers and immobilizers. Non-toxic. Relatively high abundance. Low weight. Good mechanical properties. Low abrasive nature.	Processing and functionalization requirements (cross-linking, hydrolysis, modification, autoclaving, drying). Biodegradability. Moisture absorption. Long processing time. High operational cost. Poor thermal stability. Energy extensive regeneration process. Poor compatibility with hydrophobic catalysts. Quality variations.
Starch	A wide variety of sources.	Poor mechanical properties compared to other biopolymers.
Alginate	Good sorption ability. Excellent film-forming ability. High availability.	Difficult to work with. Poor mechanical properties. Poor dimensional stability.
Guar gum	A superior swelling capacity leading to lower diffusion of contaminants in water.	Lower abundance. Poor mechanical and thermal properties. Microbial contamination. High hydration rate. Poor shear resistance.

that have been utilized in the metal, metal sulfide, and metal oxide semiconductor photocatalytic processes for the removal of dye. The important details about the recently reported biopolymers incorporated with metal, metal oxide, and metal sulfide based advanced oxidation processes (AOPs; photocatalysis) in the literature are summarized in table 5.2.

Table 5.2. Comparative summary of biopolymers incorporating metal, metal oxide, and metal sulfide nanomaterials for dye removal.

Photocatalysts		Light source	Dye	% of degradation	Reference
Metals	Chitosan/silver	Sunlight	MB	88.0%	[17]
	Se NPs	UV	MB (methylene blue)	60%	[18]
	Ag NPs/metal organic framework (MIL-100) (Fe)/guar gum	Visible light	MB	100%	[19]
	Gelatin–Zr(IV) phosphate nanocomposite	Solar light	Methylene blue, Fast Green	87.81%, 89.91%	[20]
	Ag NPs using κ-carrageenan	$NaBH_4$ as reducing agent	Rhodamine B, methylene blue	100%	[21]
	Pectin/ZSPNC	Solar irradiation	MB	97.02	[22]
Metal oxides	Chitosan–CuO nanocomposite	Sunlight during April–May 2019 (11 am to 1 pm)	CR (Congo red), MB	90.8%, 91.3%	[23]
	CS–ZnO nanocomposite	Sunlight (April–May 2018) 10 am to 1 pm	MB, CR	82%, 86%	[24]
	CS/ZnO nanocomposite	Sunlight	CR	91%	[25]
	Chitosan/ZnO fiber	Ultraviolet light	Methylene blue, methyl orange, rhodamine B	85.8%, 94.3%, 99.3%	[26]
	TiO_2/chitosan/glycerol	Sunlight	Acid blue 193	85.0%	[27]
	NiO/Arabic gum	UV	MB	82%	[28]
	TiO_2/natural gum Arabic	Visible light	MB	90%	[29]
	Fe_3O_4/tragacanth gum	Sunlight	Congo red	89.51%	[30]
	Ag_2O–guar gum	Visible light	Nitrobenzene, benzyl alcohol	96.0%, 98.0%	[31]

(*Continued*)

Table 5.2. (*Continued*)

Photocatalysts		Light source	Dye	% of degradation	Reference
	Sodium alginate/PVP/ WO_3	Visible light	Methylene blue	98%	[32]
	TiO_2/calcium alginate	Sunlight	Basic blue 41	96.0%	[33]
	Sodium alginate–TiO_2-bentonite	Ultraviolet light	Methylene blue	90.50%	[34]
	Sb_2S_3–CeO_2/chitosan–starch	Ultraviolet–visible light	Paraquat	~98%	[35]
	Bi_2WO_6–TiO_2/starch	Visible light	Ethylene	12.47%	[36]
	Cellulose/zinc oxide	Ultraviolet light	Methyl orange	99.02%	[37]
	$H_3PW_{12}O_{40}$/TiO_2/ cellulose	Ultraviolet light	Methylene blue	90%	[38]
	Cr_2O_3/cellulose	Ultraviolet light	Crystal violet	99.65%	[39]
	g-C_3N_4–green cellulose nanofibers	Visible light	Rhodamine B	95%	[40]
	Nb/starch-doped ZnO	Visible light	MB	98.1%	[41]
	Guar gum/Al_2O_3	Solar irradiation	Malachite green	90%	[42]
Metal sulfides	ZnS tethered chitosan–polyaniline	Ultraviolet–visible light	Methylene blue, malachite green	87.3%, 91.5%	[43]
	CdS-sensitized nano-ZnO/chitosan	Sunlight	Congo red	94.34%	[44]
	CdS/carboxyl methyl starch	Visible light	Methylene blue	86.72%	[45]
	Chitosan and C_3N_4-doped CdS	Visible light	MB	99.75%	[46]
	ZnS/chitosan	UV irradiation	Methylene blue	90%	[47]

5.2 Synthesis of green polymer based metal and metal oxide nanomaterials

Starch, cellulose, chitosan, and dextran are some instances of biopolymers that are utilized frequently in the production of nanomaterials or composites [3, 4]. These biopolymers can be derived from a variety of organisms, including fungi, bacteria, plants, and algae. When synthesizing metal/polymer nanocomposites, *ex situ* and *in situ* processes are utilized the majority of the time. *Ex situ* entails the independent synthesis of polymer and metal nanoparticles (NPs), followed by their combination using either the melt compounding method or the solution blending method. *Ex situ* is performed in a separate facility. The *in situ* approach, on the other hand, involves the dispersion of metal nanoparticles in a monomeric solution and the subsequent polymerization of those nanoparticles [5]. *In situ* preparation can also be

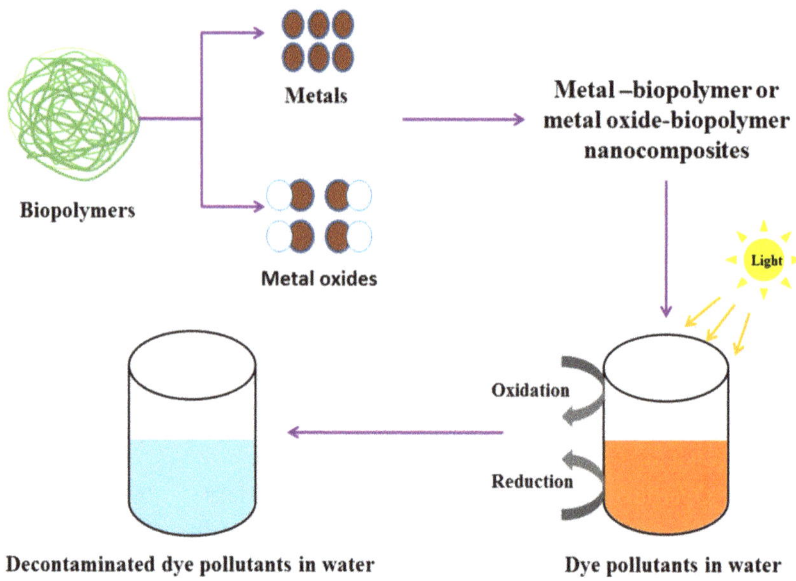

Figure 5.1. Schematic representation of dye degradation applications of metal and metal oxide incorporated biopolymer nanocomposites. (Reproduced from [8]. Creative Commons Attribution 4.0 International License http://creativecommons.org/licenses/by/4.0/.)

accomplished through the reduction of metal ions, followed by polymerization. Methods such as pulsed laser deposition, thermal evaporation, molecular beam epitaxy, sputtering deposition, solvothermal, sol–gel, and other similar techniques are utilized in the synthesis of metal oxide nanoparticles. Methods such as direct mixing or blending, the sol–gel technique, and *in situ* polymerization are used to prepare novel metal oxide/biopolymer composites [6]. Sol–gel has fewer advantages than other techniques, which have a lower sintering temperature, high purity, and more uniform particle distribution. However, each of these three methods is utilized quite frequently in the preparation procedure. Because the reaction parameters have an impact on the morphology of the composite, they must be accurately monitored during the fabrication of these composites. These parameters include temperature, the kind of solvent used, and how much of it is used. Researchers have suggested green synthesis strategies in addition to standard procedures in order to avoid the utilization of harmful chemical compounds in the synthesis process [7]. Figure 5.1 shows a schematic illustration of the applications of metal and metal oxide incorporated biopolymer nanocomposites for the degradation of dyes.

5.3 Impact of green polymers on metal nanomaterials

Metal-incorporated nanostructured materials have shown efficient photodegradation of potentially toxic dyes. This can be attributable to the pore size, optimal surface properties, and mechanical characteristics of the materials [9]. Because of the increased transfer of produced charges, metal oxidation is decreased by incorporating metal into various polymeric nanomaterials. Both the lifespan and reactivity of electrons have been extended because of this technique. It is possible that grafting metal nanoparticles onto different polymeric substances will make them more stable and compact [10]. When compared to the individual metal nanomaterials, the resulting metal-incorporated polymeric materials had a more uniform size distribution than the individual metal nanomaterials. As a consequence of this, a substantially steadier performance is attainable. In addition, nanomaterials with metal incorporation show a significant degree of effectiveness in the photodegradation of azo dyes. This is because of the optimum mechanical stiffness, pore size, and surface plane chemistry of the materials [11]. An electron transport mechanism and a redox reaction are at the heart of photocatalytic degradation's fundamental components. If the semi-conductivity is changed by including metal nanoparticles, then the system will encourage charge movement at the interface, which will, in turn, reduce the amount of oxidation that the metal experiences. The lifespan of electrons is increased due to this action, which is then followed by a rise in reactivity. Because of the small particle size, a phenomenon known as surface plasmon resonance (SPR) occurs more frequently when electrons coexist [12]. Polymeric nanomaterials that have metal incorporated into them function as a stabilizer for the material. Industrial applications have been discovered for nanomaterials combined with polymeric materials such as resin to develop reaction catalysts. It is also possible to graft metal nanoparticles onto other polymeric materials, which not only gives a distinct performance compared to metal monomeric nanoparticles but also boosts the compactness and stability of the resulting material. Enoki mushroom fruit-body extract was utilized for the very first time by Rabeea *et al* in order to create extracellular mycosynthesized triangular and spherical gold nanoparticles [13]. As-synthesized nanoparticles were used in the reaction to break down methylene blue dye. In their research, Tamayo and colleagues claimed that gold nanostructured particles have a wide substrate specificity, which gives them an exceptional capacity to break down potentially dangerous chemical dyes [10]. Previously, silver nanoparticles were incorporated using various plant parts and species extracts, including *Agaricus bisporus*, *Solanum tuberosum*, *Pleurotus sajor caju*, *Ganoderma lucidum*, *Pleurotus ostreatus*, and *Terminalia chebula*. These materials were used to study the deterioration of dyestuffs and for a number of other purposes [8]. Lanthanum was utilized in addition to silver and gold owing to its ability to interact with the functional groups found in biopolymers, which made it possible to produce complexes with a higher surface area. Meenakshi and Sirajudheen investigated a chitosan/lanthanum (La^{3+})/graphite composite to research the photodegradation of MB dye [14]. Similarly, Kusrini *et al* investigated the effects of a composite composed of modified graphite waste, lanthanum, and chitosan on various dyes, including methyl violet, rhodamine B, methyl orange, and methylene blue [15]. In particular, the author pointed out that an increased lanthanum (0.03 M) concentration led to a more significant amount of adsorption. When a polymer is incorporated into a material, the nanoparticle sizes become more uniform, which, in turn, makes the substance more stable. Typically, the sizes of metal nanoparticles vary. This was demonstrated successfully during the process of integrating bacterial cellulose fibers with Au nanoparticles. Incorporating Au nanoparticles with a size of

74.32 nm into biomaterials obtained from the fresh fruiting bodies of the Enoki mushroom caused the degradation of over 75% of the MB dye in just four hours [13]. Because they have such a broad substrate specificity, gold nanoparticles can treat a wide variety of organic dyes [16]. In the case of lanthanum (La), the f-orbital electron of the metal ions interacts with various functional groups present in different biopolymers to produce complexes with higher surface areas. Incorporating lanthanum into biopolymers leads to increased photocatalytic activity along with improved adsorption capacity, particularly for organic chemicals such as azo dyes. A chitosan and lanthanum metal combination destroyed 90% of methylene blue in just forty minutes. In this location, the chemical oxygen demand (COD) dropped by about eight times, indicative of the mineralization of methylene blue [14].

5.4 Impact of green polymers on metal oxide nanomaterials

Metal oxides have been shown to be powerful photocatalysts when doped with biopolymers, in addition to the incorporation of metals. The physical and chemical characteristics of the resulting nanomaterials are enhanced when biopolymers are combined with metal oxides. This, in turn, leads to an improvement in the ability of the nanomaterials to decompose the intricate chemical structure of toxic dyes, which resulted in a higher level of photocatalytic efficiency [48]. Because of ZnO's stability, semi-conductivity, and activity, it is one of the most effective nanomaterials for photocatalytic dye degradation [49]. It has been demonstrated that a zinc oxide nanomaterial based on starch possesses superior conductivity and dielectric characteristics compared to nanoparticles of pure metal oxide [49]. Considering their higher thermal stability than pure metal oxide, ZnO nanoparticles made from cellulose can be used for extensive wastewater treatment, including dyes [50]. Utilizing reactive red 2 dyes, researchers investigated the degree to which ZnO-incorporated chitosan beads were able to degrade the dye over time [51]. Under visible and UV light irradiation, the researchers evaluated both the individual ZnO beads as well as the ZnO/chitosan beads. According to observations, ZnO/chitosan beads and ZnO under visible light and UV radiation, respectively, displayed remarkable advantages. The highest efficiency reactive red 2 dye decomposition was accomplished when the ZnO catalyst and ultraviolet light were used. Additionally, Haldorai and Shim investigated the performance of methylene blue dye degradation by synthesizing chitosan-encased ZnO [52]. The researcher recycled the ZnO/chitosan hybrid photocatalyst five times in this study to assess its impact on the dye's photodegradation performance. However, the results did not reveal any significant modification. Azo dyes, including methyl orange, methylene blue, and Congo red, have been successfully treated with ZnO/carbon black grafted in cellulose acetate [53]. Khan *et al* studied the degrading impact of a CuAg/ZnO/carbon black/cellulose acetate hybrid nanocomposite on several hazardous dyes such as Congo red, rhodamine B, methyl orange, and methylene blue. Of these dyes, Congo red showed the quickest disintegrating rate [53]. In addition to ZnO/biopolymer composites, results from TiO_2-incorporated biopolymers have also suggested that workable dye degradation can occur. Dassanayake and colleagues reported the synthesis of aerochitin/TiO_2, as well as its application in the breakdown of methylene blue dye [54]. Because aerochitin is such a good sorbent, the addition of TiO_2 led to an increase in its photocatalytic activity. Composites made of TiO_2 and chitosan are another area of study. The incorporation of TiO_2 into chitosan led to an increase in the latter substance's catalytic activity as well as facilitated its recovery. To investigate the degradation of three distinct dyes (reactive red 2, methylene blue, and rhodamine B), Farzana and Meenakshi *et al* created a TiO_2/chitosan composite material. When compared to the other two dyes, it was discovered that reactive red 2 achieved nearly total breakdown when exposed to UV light [55]. Bahal *et al* produced TiO_2/chitosan/acrylic acid and used it to break down malachite green dye while the mixture was being irradiated with visible light to obtain a further increase in the surface characteristics of the material [56]. The incorporation of chitosan resulted in the production of a robust polymeric backbone, which assisted the functional groups in effectively degrading the dye. It was claimed that a chitosan–acrylic acid biopolymer grafted with nano-TiO_2 could dissolve more than 90% of the malachite green that was present in wastewater while maintaining a neutral pH. This was accomplished by a visible light-mediated photocatalytic method [56]. In addition to aerochitin and chitosan, cellulose and TiO_2 composites were also advised by researchers as a means of improving photodegradability [8]. To study the breakdown of methylene blue dye, Han *et al* created a TiO_2/hydroxyethyl cellulose/carboxymethyl hydrogel [57]. Not only did the as-synthesized composite improve the contact area, but it also made the possibility of the catalyst and the dye interacting more likely. The composite demonstrated, in addition to improved photocatalytic efficiency, regeneration capabilities as well as self-cleaning qualities. After being modified with nano-TiO_2, cyclodextrin, which is an oligosaccharide, is utilized for wastewater treatment. Cyclodextrin is formed from the enzymatic conversion of starch [58]. In addition to zinc oxide and titanium dioxide, other metal oxides, such as iron oxide, silver oxide, and manganese dioxide, that are incorporated into biocomposites have also been investigated for use in polluted water treatment applications; however, this field of work still calls for additional research using a more comprehensive strategy [48, 59, 60]. Chitosan with nano-iron oxide (Fe_3O_4) has been utilized to filtrate wastewater containing dye [48]. Due to magnetic dipole–dipole interactions during the breakdown of the dangerous X-3B dye, chitosan/nanoFe_3O_4 nanomaterial is also gaining attention [59]. Few bio-sourced enzymes destroy reactive azo dyes. Laccase has a significant amount of promise for use in the bioremediation of dye-containing wastewater [61]. Within 20 min, the azo dye procion red MX-5B was practically completely degraded thanks to nano-Fe_3O_4/SiO_2 supported by immobilized laccase [61]. Nanoparticles of MnO_2/cellulose with a size of less than 100 nm destroyed 90% of indigo carmine after 25 min when exposed to natural light and an acidic pH [60]. In this application, the biopolymeric nanoparticles may be extracted from the solution and recycled a minimum of ten times without causing any degradation to the decolorization performance. This proves that, in the presence of photons, metal-oxide-incorporated biopolymeric nanomaterials break down complex azo dyes in a relatively short

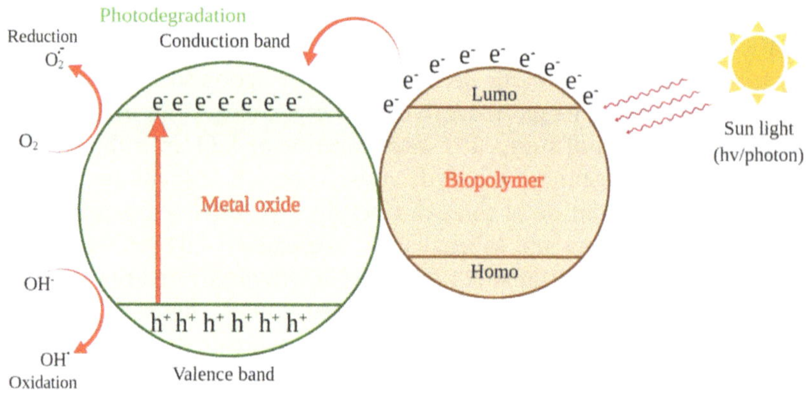

Figure 5.2. Mechanism of photocatalytic dye degradation using green polymer doped metal oxide nanomaterials. (Reproduced with permission from [62]. Copyright 2020 Springer Nature.)

amount of time through a process that is safe for the environment and they can be recycled. Because of this, incorporating biopolymers into metal or metal oxide not only improved the breakdown of contaminants but also provided recyclability and a safe method for the environment. Figure 5.2 shows a schematic illustration of the photodegradation of dyes using green polymer incorporated metal oxide nanomaterials

5.5 Impact of green polymers on metal sulfide nanomaterials

The biopolymer aids in the crystallization of ZnS nanoparticles [63]. ZnS supported by a cellulose/chitosan sponge of 15 nm size demonstrated effective photocatalytic activity and was utilized in the process of photodegradation of Congo red in wastewater [64]. As a result of the integration of cellulose/chitosan, which was 96.53%, the degradation rate was accelerated. The possible mechanism for degradation of the organic dye methylene blue using cellulose/chitosan sponge supported by ZnS is depicted in figure 5.3. According to the findings, not only do polysaccharide substrates have a strong ability to adsorb, but they also supply a significant number of hydroxyl groups on their surfaces. These hydroxyl groups have the potential to react with the holes, which would result in an increase in the generation of radical ˙OH. These radicals of ˙OH are able to adequately degrade the organic dyes that have been adsorbed onto the composite surface. The dye molecules are also broken down into their constituent parts by the active substances radical $^{\cdot}O_2^-$ and radical ˙OH.

Metal sulfide nanomaterials are synthesized using the elements lead (Pb) and cadmium (Cd). For instance, the tiny CdS materials that *Klebsiella pneumonia* can make on its cell membrane are electron-dense. These substances might encourage methyl orange's photoreduction through a consequent electron transfer [65]. The elimination of potentially harmful dyes has been accomplished with nano-chitosan. Methylene blue may be removed using ZnS, nano-photocatalysts based on chitosan,

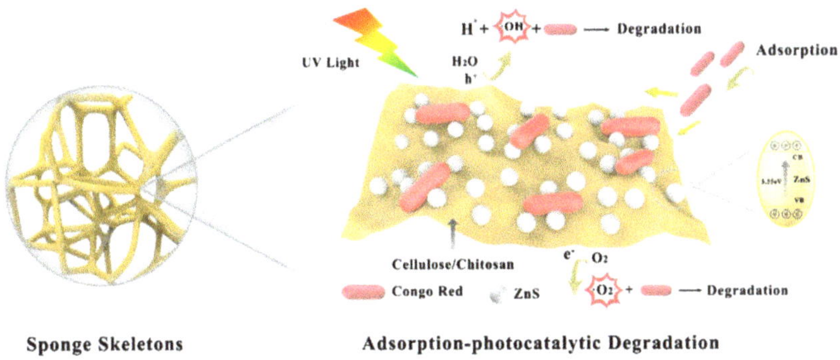

Figure 5.3. Photocatalytic degradation mechanism of ZnS-supported by cellulose/chitosan sponge. (Reproduced with permission from [64]. Copyright 2022 Elsevier.)

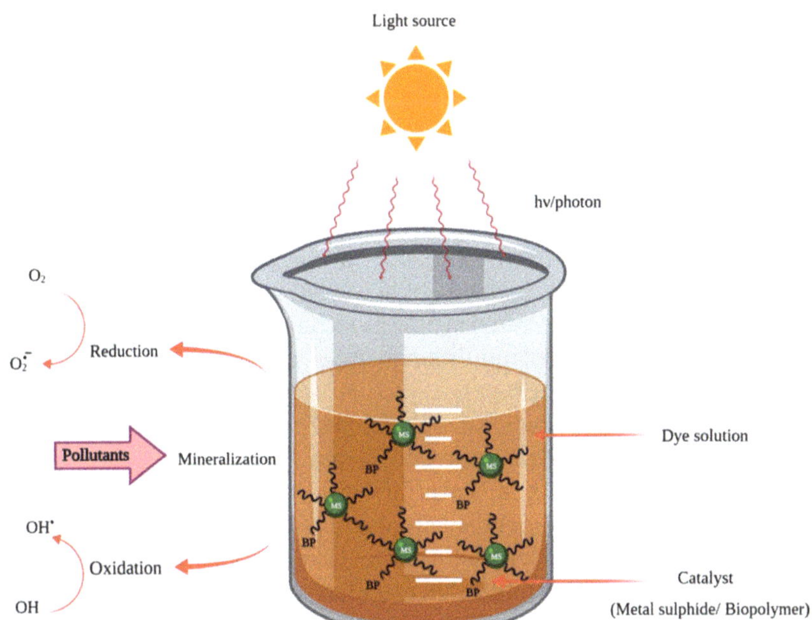

Figure 5.4. Rationale behind the synthesis of metal sulfide-incorporated biopolymeric nanoparticles as photocatalysts for dye degradation. BP: biopolymer; MS: metal sulfide. (Reproduced with permission from [62]. Copyright 2020 Springer Nature.)

and quantum dots made from chitosan [47]. In this context, ZnS performs the role of a semiconductor, which increases the efficiency of the removal. Within 90 min, this ZnS/chitosan-based nanomaterial may generate an almost 3.5 nm in size photo-reduction of MB by oxidation when exposed to UV irradiation. Figure 5.4 shows the production of biopolymeric nanoparticles with metal sulfide incorporation.

5.6 Modified/engineered nanomaterials for water treatment

The photocatalyst approach is believed to have a number of serious flaws, one of the most significant of which is the quick recombination that occurs when TiO_2 is used. Because of the phenomenon of electron recombination, the quantum efficiency of TiO_2's photocatalytic properties is reduced with time. In addition, its application in sunlight is limited because it does not absorb visible light. Numerous investigations into the manipulation of TiO_2 properties have been carried out to boost the photoactivity of TiO_2. The absorption of titanium dioxide may be modulated, as proven by Hoffman *et al*, by substituting metallic atoms for titanium atoms. This occurs due to the introduction of localized energy levels between the valance band (VB) and the conduction band (CB) of the compound [66]. In the photocatalyst approach, the energy levels present at each level, as well as the dose and distribution of dopants inside the semiconductors, serve crucial purposes. One way to overcome the limitations of titanium dioxide photocatalyst is to dope it with various metals such as gold, copper, chromium, cobalt, and iron. In spite of the fact that metal doping has a beneficial effect on the efficiency of photocatalysis, a number of investigations have demonstrated that it can result in recombination, which is detrimental to photocatalytic efficiency [67, 68]. Nanoparticles of various metals, including platinum, silver, and palladium, have been used to modify the properties of TiO_2. The findings of a number of different experimental investigations carried out by various researchers indicate that the surface modification of metal nanoparticles may improve the photocatalytic activity of TiO_2 in visible light. In comparison to other entire noble metals, silver's antibacterial properties, ease of manufacture, and relatively low cost give it the greatest potential for widespread application. Electrospinning, hydrothermal, ultraviolet radiation, and chemical reduction are only a few of the many synthesis techniques that have been thoroughly investigated so far for synthesizing different types of Ag/TiO_2 nanostructures that may be used in photocatalytic applications. According to their findings, Ag can significantly reduce the recombination of holes and electrons in titanium dioxide, but it also causes a redshift in the wavelength of titanium dioxide's absorption edge. As a result, the photocatalytic activity in visible radiation is improved to some degree [69]. Kappadan *et al* conducted research on the potential of tetragonal $BaTiO_3$ nanoparticles as efficient photocatalytic materials for the breakdown of organic pollutants [70]. According to their findings, the photoactivity of tetragonal $BaTiO_3$ was much higher when compared with that of cubic $BaTiO_3$. In addition, the increased photoactivity of tetragonal $BaTiO_3$ was demonstrated by the photoluminescence results. Tetragonal $BaTiO_3$ has an increased photoactivity, making it a competitive photocatalytic material to break down pollutants [70]. In a separate piece of research, a $BaTiO_3/ZnO$ heterostructure photocatalyst that was shown to be more efficient in the mineralization of dye was investigated. It was created using the hydrothermal method, during which $BaTiO_3$ nanoparticles were attached to ZnO that was fashioned like hexagonal rods. The photocatalytic degradation of MB dye was carried out under ultraviolet light in order to assess the efficacy of the $BaTiO_3/ZnO$ heterostructure as a photocatalyst. When compared to $BaTiO_3$, the photoactivity of the $BaTiO_3/ZnO$ heterostructure was found to be much higher. This may be explained by the effective charge carrier separation that occurred due to the development of a heterojunction between $BaTiO_3$ and ZnO. Within one hour of being exposed to ultraviolet radiation, about 94% of MB was broken down into its constituent parts. They concluded that the $BaTiO_3/ZnO$ heterostructure might be a competitive alternative for wastewater treatment [71]. Although metal doping of titanium dioxide has been shown to be effective for photocatalytic activity when exposed to visible light, the process is not without its drawbacks. These include the potential for gas poisoning, the high cost of the process, and the adverse environmental effects that it can have. In most cases, the desire for higher economic feasibility in industrial-scale applications results in the replacement of such elements with transition or non-metal components that are more cost-effective. Because of the difficulties associated with metal-based photocatalysts, the modification of these photocatalysts with non-metal components was researched in depth to create visible light active stable TiO_2. By generating intermediate energy levels between the CB and the VB, or by reducing the amount of energy in the bandgap, this approach decreased the excitation energy of titanium dioxide.

5.7 Challenges of nanomaterials for water treatment

In wastewater treatment, the burgeoning subject of nanomaterial research is now facing a few obstacles. Nanomaterial opens up a variety of avenues for wastewater treatment, and the compounds it contains are of varying types, each of which may be differentiated on the basis of the morphology of the particle. Both the research and development of nanomaterials for use in commercial applications, as well as the manufacture of nanomaterials themselves, are advancing at a breakneck pace globally. Nanomaterials are used in various water purification processes, including photocatalytic, adsorption, and nano sorbent, to remove pollutants from contaminated water. To improve the efficiency of these approaches, various adjustments are required. In addition, our comprehension of the dangers posed by nanotechnology has not advanced at the same rate as the research that has opened the door to a variety of uses for nanomaterials. The lack of understanding of nanomaterials, including how they are released into the environment, how they migrate into the water, and how they begin to live in water, is the most significant obstacle that must be overcome. Another difficulty is associated with the state of human health because certain kinds of materials have some negative impacts or toxicological effects. Studies have been performed that indicate that nanoparticles may cause certain health problems, but researchers are currently looking deeper into this issue. It is not simple to determine toxicity and current studies have to include some trials carried out at the right doses as well as more information regarding toxicological investigations. The membrane process is likewise an excellent method for treating wastewater, although relatively few studies have been reported. Despite this, the membrane process is widely employed. The employment of the membrane in the fouling process and water treatment presents the most significant problem since, following the completion of work or filtering, the pores of the membrane might become blocked, causing its efficiency to begin to decline. In the case of nanosorbents, the reusability is the primary disadvantage; hence, it is necessary to synthesize such types of nanomaterials as are capable of effectively removing the contaminants from wastewater and, after that, the recovery process of these nanosorbents should be very simple. The United States Environmental Protection Agency (USEPA) discovered that several fundamental obstacles must be overcome to remove nanoparticles from water utilizing a treatment technique [72]. (a) How do you get rid of nanoparticles in wastewater, and what mechanism do you use? (b) During wastewater treatment, what impact do nanoparticles have on the other waste compounds being removed? (c) The mechanism behind the efficient operation of various technologies, such as coagulation and carbon adsorption, among others. In addition, water is known to carry a variety of contaminants, many of which, when subjected to the processes described above, are either neutralized or broken down. It is necessary to devise an experiment in order to identify the products that are created throughout the intermediate stages of these processes. In addition, there is a requirement for ongoing advancement in techniques in order to assess nanomaterials at a low cost. The methodology also has to be suitable for evaluating complicated nanomaterials.

References

[1] Peramune D, Manatunga D C, Dassanayake R S, Premalal V, Liyanage R N, Gunathilake C and Abidi N 2022 Recent advances in biopolymer-based advanced oxidation processes for dye removal applications: a review *Environ. Res.* **215** 114242

[2] Ogundare S A and van Zyl W E 2019 A review of cellulose-based substrates for SERS: fundamentals, design principles, applications *Cellulose* **26** 6489–528

[3] Kolangare I M *et al* 2019 Antibiofouling hollow-fiber membranes for dye rejection by embedding chitosan and silver-loaded chitosan nanoparticles *Environ. Chem. Lett.* **17** 581–7

[4] Farshchi E, Pirsa S, Roufegarinejad L, Alizadeh M and Rezazad M 2019 Photocatalytic/biodegradable film based on carboxymethyl cellulose, modified by gelatin and TiO_2–Ag nanoparticles *Carbohydr. Polym.* **216** 189–96

[5] Tamayo L, Palza H, Bejarano J and Zapata P A 2018 Polymer composites with metal nanoparticles: synthesis, properties, and applications. synthesis, properties, and applications *Polymer Composites with Functionalized Nanoparticles: Synthesis, Properties, and Applications* (Amsterdam: Elsevier) pp 249–86

[6] Prasanna S R V S, Pandey S, Balaji K and Rana S 2019 Metal oxide based nanomaterials and their polymer nanocomposites *Nanomaterials and Polymer Nanocomposites: Raw Materials to Application* (Amsterdam: Elsevier) pp 123–44

[7] Edison T J I and Sethuraman M G 2012 Instant green synthesis of silver nanoparticles using *Terminalia chebula* fruit extract and evaluation of their catalytic activity on reduction of methylene blue *Process Biochem.* **47** 1351–7

[8] Dutta S, Pohrmen C B, Banerjee I, Trivedi A, Verma R and Dubey S 2021 Bio-derived metal and metal oxide incorporated biopolymer nanocomposites for dye degradation applications: a review *Octa J. Biosci.* **9** 30–6

[9] Opoku F, Kiarii E M, Govender P P and Mamo M A 2017 Metal oxide polymer nanocomposites in water treatments *Descriptive Inorganic Chemistry Researches of Metal Compounds* (London: InTech Open)

[10] Tamayo L, Palza H, Bejarano J and Zapata P A 2018 Polymer composites with metal nanoparticles: synthesis, properties, and applications. synthesis, properties, and applications *Polymer Composites with Functionalized Nanoparticles: Synthesis, Properties, and Applications* (Amsterdam: Elsevier) pp 249–86

[11] Opoku F, Kiarii E M, Govender P P and Mamo M A 2017 Metal oxide polymer nanocomposites in water treatments *Descriptive Inorganic Chemistry Researches of Metal Compounds* (London: Intech Open)

[12] Sankar R, Rizwana K, Shivashangari K S and Ravikumar V 2015 Ultra-rapid photocatalytic activity of *Azadirachta indica* engineered colloidal titanium dioxide nanoparticles *Appl. Nanosci.* **5** 731–6

[13] Rabeea M A, Owaid M N, Aziz A A, Jameel M S and Dheyab M A 2020 Mycosynthesis of gold nanoparticles using the extract of *Flammulina velutipes*, Physalacriaceae, and their efficacy for decolorization of methylene blue *J. Environ. Chem. Eng.* **8** 103841

[14] Sirajudheen P and Meenakshi S 2019 Facile synthesis of chitosan–La^{3+}–graphite composite and its influence in photocatalytic degradation of methylene blue *Int. J. Biol. Macromol.* **133** 253–61

[15] Kusrini E, Wicaksono B, Yulizar Y, Prasetyanto E A and Gunawan C 2018 Textile dye removal from aqueous solution using modified graphite waste/lanthanum/chitosan composite *IOP Conf. Ser.: Mater. Sci. Eng.* **316** 012029

[16] Pielichowski K and Majka T M 2018 *Polymer Composites with Functionalized Nanoparticles: Synthesis, Properties, And Applications* (Amsterdam: Elsevier)

[17] Nandana C N, Christeena M and Bharathi D 2022 Synthesis and characterization of chitosan/silver nanocomposite using rutin for antibacterial, antioxidant and photocatalytic applications *J. Clust. Sci.* **33** 269–79

[18] Kazemi M, Akbari A, Sabouri Z, Soleimanpour S, Zarrinfar H, Khatami M and Darroudi M 2021 Green synthesis of colloidal selenium nanoparticles in starch solutions and investigation of their photocatalytic, antimicrobial, and cytotoxicity effects *Bioprocess. Biosyst. Eng.* **44** 1215–25

[19] Duan C *et al* 2020 Facile synthesis of Ag NPs@ MIL-100(Fe)/guar gum hybrid hydrogel as a versatile photocatalyst for wastewater remediation: photocatalytic degradation, water/oil separation and bacterial inactivation *Carbohydr. Polym.* **230** 115642

[20] Thakur M, Sharma G, Ahamad T, Ghfar A A, Pathania D and Naushad M 2017 Efficient photocatalytic degradation of toxic dyes from aqueous environment using gelatin-Zr(IV) phosphate nanocomposite and its antimicrobial activity *Colloids Surf.* B **157** 456–63

[21] Pandey S, Do J Y, Kim J and Kang M 2020 Fast and highly efficient catalytic degradation of dyes using κ-carrageenan stabilized silver nanoparticles nanocatalyst *Carbohydr. Polym* **230** 115597

[22] Pathania D, Sharma G and Thakur R 2015 Pectin @ zirconium (IV) silicophosphate nanocomposite ion exchanger: photo catalysis, heavy metal separation and antibacterial activity *Chem. Eng. J.* **267** 235–44

[23] Sathiyavimal S, Vasantharaj S, Kaliannan T and Pugazhendhi A 2020 Eco-biocompatibility of chitosan coated biosynthesized copper oxide nanocomposite for enhanced industrial (azo) dye removal from aqueous solution and antibacterial properties *Carbohydr. Polym.* **241** 116243

[24] Bharathi D, Ranjithkumar R, Chandarshekar B and Bhuvaneshwari V 2019 Preparation of chitosan coated zinc oxide nanocomposite for enhanced antibacterial and photocatalytic activity: as a bionanocomposite *Int. J. Biol. Macromol.* **129** 989–96

[25] Madhan G, Begam A A, Varsha L V, Ranjithkumar R and Bharathi D 2021 Facile synthesis and characterization of chitosan/zinc oxide nanocomposite for enhanced antibacterial and photocatalytic activity *Int. J. Biol. Macromol.* **190** 259–69

[26] Li H, Hao M X, Kang H R and Chu L Q 2021 Facile production of three-dimensional chitosan fiber embedded with zinc oxide as recoverable photocatalyst for organic dye degradation *Int. J. Biol. Macromol.* **181** 150–9

[27] Hoang N T T, Tran A T K, Hoang M H, Nguyen T T H and Bui X T 2021 Synergistic effect of TiO_2/chitosan/glycerol photocatalyst on color and COD removal from a dyeing and textile secondary effluent *Environ. Technol. Innov.* **21** 101255

[28] Sabouri Z, Akbari A, Hosseini H A, Khatami M and Darroudi M 2021 Green-based bio-synthesis of nickel oxide nanoparticles in Arabic gum and examination of their cytotoxicity, photocatalytic and antibacterial effects *Green Chem. Lett. Rev.* **14** 402–12

[29] Araujo F P, Honorio L M C, Lima I S, Trigueiro P, Almeida L C, Fechine P B A, Santos F E P, Peña-Garcia R, Silva-Filho E C and Osajima J A 2020 New composite TiO_2/naturals gums for high efficiency in photodiscoloration process *Ceram. Int.* **46** 15534–43

[30] Sharma A, Pathania D and Kumar A 2020 Bio-polymer based tragacanth gum (TG) loaded Fe_3O_4 nanocomposite for the sequestration of tenacious Congo red dye from waste water *J. Mater. Sci. Technol. Res* **7** 92–100

[31] Balachandramohan J, Sivasankar T and Sivakumar M 2020 Facile sonochemical synthesis of Ag_2O–guar gum nanocomposite as a visible light photocatalyst for the organic transformation reactions *J. Hazard. Mater.* **385** 121621

[32] Elsayed E M, Elnouby M S, Gouda S M H, Elessawy N A and Santos D M F 2020 Effect of the morphology of tungsten oxide embedded in sodium alginate/polyvinylpyrrolidone composite beads on the photocatalytic degradation of methylene blue dye solution *Materials* **13** 1905

[33] Nouri L, Hemidouche S, Boudjemaa A, Kaouah F, Sadaoui Z and Bachari K 2020 Elaboration and characterization of photobiocomposite beads, based on titanium (IV) oxide and sodium alginate biopolymer, for basic blue 41 adsorption/photocatalytic degradation *Int. J. Biol. Macromol.* **151** 66–84

[34] Subekti R and Helmiyati H 2020 Sodium alginate–TiO_2–bentonite nanocomposite synthesis for photocatalysis of methylene blue dye removal *IOP Conf. Ser.: Mater. Sci. Eng.* **763** 012018

[35] Hosseini M, Pourabadeh A, Fakhri A, Hallajzadeh J and Tahami S 2018 Synthesis and characterization of Sb_2S_3–CeO_2/chitosan–starch as a heterojunction catalyst for photo-degradation of toxic herbicide compound: optical, photo-reusable, antibacterial and anti-fungal performances *Int. J. Biol. Macromol.* **118** 2108–12

[36] Wang H, Wang L, Ye S and Song X 2019 Construction of Bi_2WO_6–TiO_2/starch nano-composite films for visible-light catalytic degradation of ethylene *Food Hydrocoll.* **88** 92–100

[37] Hasanpour M, Motahari S, Jing D and Hatami M 2021 Investigation of operation parameters on the removal efficiency of methyl orange pollutant by cellulose/zinc oxide hybrid aerogel *Chemosphere* **284** 131320

[38] Lin Z and Huang J 2021 A hierarchical $H_3PW_{12}O_{40}$/TiO_2 nanocomposite with cellulose as scaffold for photocatalytic degradation of organic pollutants *Sep. Purif. Technol.* **264** 118427

[39] Lu M, Cui Y, Zhao S and Fakhri A 2020 Cr_2O_3/cellulose hybrid nanocomposites with unique properties: facile synthesis, photocatalytic, bactericidal and antioxidant application *J. Photochem. Photobiol.* B **205** 111842

[40] Anusuyadevi P R, Riazanova A V, Hedenqvist M S and Svagan A J 2020 Floating photocatalysts for effluent refinement based on stable pickering cellulose foams and graphitic carbon nitride (g-C_3N_4) *ACS Omega* **5** 22411–9

[41] Ikram M, Shahid H, Haider J, Haider A, Naz S, Ul-Hamid A, Shahzadi I, Naz M, Nabgan W and Ali S 2022 Nb/starch-doped ZnO nanostructures for polluted water treatment and antimicrobial applications: molecular docking analysis *ACS Omega* **7** 39347–61

[42] Pathania D, Katwal R, Sharma G, Naushad M, Khan M R and Al-Muhtaseb A H 2016 Novel guar gum/Al_2O_3 nanocomposite as an effective photocatalyst for the degradation of malachite green dye *Int. J. Biol. Macromol.* **87** 366–74

[43] Sirajudheen P, Resha Kasim V C, Nabeena C P, Basheer M C and Meenakshi S 2021 Tunable photocatalytic oxidation response of ZnS tethered chitosan–polyaniline composite for the removal of organic pollutants: a mechanistic perspective *Mater. Today Proc.* **47** 2553–9

[44] Jiang R, Zhu H Y, Fu Y Q, Jiang S T, Zong E M, Zhu J Q, Zhu Y Y and Chen L F 2021 Colloidal CdS sensitized nano-ZnO/chitosan hydrogel with fast and efficient photocatalytic removal of Congo red under solar light irradiation *Int. J. Biol. Macromol.* **174** 52–60

[45] Xue Y, Chang Q, Hu X, Cai J and Yang H 2020 A simple strategy for selective photocatalysis degradation of organic dyes through selective adsorption enrichment by using a complex film of CdS and carboxylmethyl starch *J. Environ. Manage.* **274** 111184

[46] Ikram M, Maghfoor H, Shahzadi A, Haider A, Shahzadi I, Abid N, Ul-Hamid A, Haider J, Nabgan W and Butt A R 2022 Towards effective dye degradation and antimicrobial behavior of chitosan and C_3N_4-doped CdS nanoparticles *Mater. Today Commun.* **33** 104814

[47] Mansur H S and Mansur A A P 2015 Nano-photocatalysts based on ZnS quantum dots/chitosan for the photodegradation of dye pollutants *IOP Conf. Ser.: Mater. Sci. Eng.* **76** 012003

[48] Prasanna S R V S, Pandey S, Balaji K and Rana S 2019 Metal oxide based nanomaterials and their polymer nanocomposites *Nanomaterials and Polymer Nanocomposites Raw Materials to Application* (Amsterdam: Elsevier) pp 123–44

[49] Ravishankar T N, Manjunatha K, Ramakrishnappa T, Nagaraju G, Kumar D, Sarakar S, Anandakumar B S, Chandrappa G T, Reddy V and Dupont J 2014 Comparison of the photocatalytic degradation of trypan blue by undoped and silver-doped zinc oxide nanoparticles *Mater. Sci. Semicond. Process.* **26** 7–17

[50] Azizi S, Ahmad M B, Hussein M Z and Ibrahim N A 2013 Synthesis, antibacterial and thermal studies of cellulose nanocrystal stabilized ZnO–Ag heterostructure nanoparticles *Molecules* **18** 6269–80

[51] Farzana M H and Meenakshi S 2015 Exploitation of zinc oxide impregnated chitosan beads for the photocatalytic decolorization of an azo dye *Int. J. Biol. Macromol.* **72** 900–10

[52] Haldorai Y and Shim J J 2013 Chitosan–zinc oxide hybrid composite for enhanced dye degradation and antibacterial activity *Compos. Interfaces* **20** 365–77

[53] Khan S A, Khan S B, Farooq A and Asiri A M 2019 A facile synthesis of CuAg nanoparticles on highly porous ZnO/carbon black-cellulose acetate sheets for nitroarene and azo dyes reduction/degradation *Int. J. Biol. Macromol.* **130** 288–99

[54] Dassanayake R S, Rajakaruna E and Abidi N 2018 Preparation of aerochitin–TiO_2 composite for efficient photocatalytic degradation of methylene blue *J. Appl. Polym. Sci.* **135** 45908

[55] Hasmath Farzana M and Meenakshi S 2014 Synergistic effect of chitosan and titanium dioxide on the removal of toxic dyes by the photodegradation technique *Ind. Eng. Chem. Res.* **53** 55–63

[56] Bahal M, Kaur N, Sharotri N and Sud D 2019 Investigations on amphoteric chitosan/TiO_2 bionanocomposites for application in visible light induced photocatalytic degradation *Adv. Polym. Technol.* **2019** 2345631

[57] Han S, Wang T and Li B 2017 Preparation of a hydroxyethyl–titanium dioxide–carboxymethyl cellulose hydrogel cage and its effect on the removal of methylene blue *J. Appl. Polym. Sci.* **134** 44925

[58] Khaoulani S, Chaker H, Cadet C, Bychkov E, Cherif L, Bengueddach A and Fourmentin S 2015 Wastewater treatment by cyclodextrin polymers and noble metal/mesoporous TiO_2 photocatalysts *Comptes Rendus Chim.* **18** 23–31

[59] Mohd Adnan M A, Bao L P and Muhd Julkapli N 2020 Mitigation of pollutants by chitosan/metallic oxide photocatalyst: a review *J. Clean. Prod.* **261** 121190

[60] Oliveira L V F, Bennici S, Josien L, Limousy L, Bizeto M A and Camilo F F 2020 Free-standing cellulose film containing manganese dioxide nanoparticles and its use in discoloration of indigo carmine dye *Carbohydr. Polym.* **230** 115621

[61] Wang H, Zhang W, Zhao J, Xu L, Zhou C, Chang L and Wang L 2013 Rapid decolorization of phenolic azo dyes by immobilized laccase with Fe_3O_4/SiO_2 nanoparticles as support *Ind. Eng. Chem. Res.* **52** 4401–7

[62] Sarkar S, Ponce N T, Banerjee A, Bandopadhyay R, Rajendran S and Lichtfouse E 2020 Green polymeric nanomaterials for the photocatalytic degradation of dyes: a review *Environ. Chem. Lett.* **18** 1569–80

[63] Tiwari A and Dhoble S J 2016 Stabilization of ZnS nanoparticles by polymeric matrices: syntheses, optical properties and recent applications *RSC Adv.* **6** 64400–20

[64] You J, Liu C, Feng X, Lu B, Xia L and Zhuang X 2022 *In situ* synthesis of ZnS nanoparticles onto cellulose/chitosan sponge for adsorption–photocatalytic removal of Congo red *Carbohydr. Polym.* **288** 119332

[65] Das R K, Pachapur V L, Lonappan L, Naghdi M, Pulicharla R, Maiti S, Cledon M, Dalila L M A, Sarma S J and Brar S K 2017 Biological synthesis of metallic nanoparticles: plants, animals and microbial aspects *Nanotechnol. Environ. Eng.* **2** 18

[66] Faraldos M and Bahamonde A 2017 Environmental applications of titania–graphene photocatalysts *Catal. Today* **285** 13–28

[67] Yu S *et al* 2021 Recent advances in metal–organic framework membranes for water treatment: a review *Sci. Total Environ.* **800** 149662

[68] Yu S, Tang H, Zhang D, Wang S, Qiu M, Song G, Fu D, Hu B and Wang X 2022 MXenes as emerging nanomaterials in water purification and environmental remediation *Sci. Total Environ.* **811** 152280

[69] Nasr M, Eid C, Habchi R, Miele P and Bechelany M 2018 Recent progress on titanium dioxide nanomaterials for photocatalytic applications *Chem. Sus. Chem.* **11** 3023–47

[70] Kappadan S, Gebreab T W, Thomas S and Kalarikkal N 2016 Tetragonal $BaTiO_3$ nanoparticles: an efficient photocatalyst for the degradation of organic pollutants *Mater. Sci. Semicond. Process.* **51** 42–7

[71] Kappadan S, Thomas S and Kalarikkal N 2020 $BaTiO_3/ZnO$ heterostructured photocatalyst with improved efficiency in dye degradation *Mater. Chem. Phys.* **255** 123583

[72] Yaqoob A A, Parveen T, Umar K and Ibrahim M N M 2020 Role of nanomaterials in the treatment of wastewater: a review *Water* **12** 495

Chapter 6

Conclusion, challenges, and future perspectives

The most significant cause of water pollution is the introduction of harmful substances such as dyes, heavy metals, pharmaceutical and personal care items, agricultural chemicals, and pesticides into fresh water supplies. These effluents entering the wastewater system lowers the quality of the soil, disrupts the process of photosynthesis, and causes the death of plant cells. In addition, when it comes to living things, water pollution can cause a variety of diseases, including intellectual disability and harm to the nervous systems of humans, as well as the loss of biodiversity, mutagenic sickness, and cancerous disease in animals. In addition to having an impact on living things, water pollution poses a significant threat to the surrounding ecosystem by, for example, making it more hazardous for life and increasing both the chemical oxygen demand (COD) and biological oxygen demand (BOD) levels. The many water treatment technologies which are described at the beginning of this book are distinct from one another in terms of the principles upon which they are based, the range of applications for which they are suitable, the rate at which they operate, and their cost. The construction, ongoing maintenance, and operating costs of any water recycling system will determine whether or not it is feasible to implement on a commercial scale. In addition to that, the management of sludge is an essential component to think about when choosing a technique. Adsorption is considered the most versatile and effective method since it can eliminate a wide range of pollutants, both inorganic and organic. When all of these considerations are taken into account, adsorption emerges as the obvious front-runner. In addition to this, it is a quick procedure that has low costs associated with its development, maintenance, and operation. In spite of the relatively high cost of installation and maintenance, reverse osmosis is a technique that sees widespread application and popularity due to the high quality of water produced by the process. Other methods, such as ultra- and microfiltration, electrodialysis, ion exchange, distillation, solvent extraction, and so on, are utilized for particular tasks, but their

doi:10.1088/978-0-7503-5517-9ch6

application in the production of drinkable and industrial levels of water is limited. The expenditures related to the tertiary, secondary, and primary processes are included in the total cost of treating the water; hence, there is a vital need to hyphenate all of these processes inexpensively to lower the total cost of treating the water.

Conventional treatment methods are unable to totally eliminate organic pollutants from wastewater. Instead, these methods produce secondary pollutants, which can often be significantly more toxic than primary pollutants. These secondary pollutants require additional treatment, which drives up the overall cost of the process. The application of advanced oxidation processes (AOPs) will allow for the resolution of these issues. Among the AOPs, photodegradation (PD) is a sustainable and energy-saving method for the treatment of wastewater. It also offers significant promise for increasing the degradation of organic pollutants that are resistant to degradation. As it removes colors from wastewater without producing any hazardous waste, this treatment method is hailed as an environmentally friendly option. The formation of active $^{\cdot}OH$ and $^{\cdot}O_2^-$ radicals is the basis for this process. These radicals totally mineralize dyes into more straightforward and nontoxic molecules, such as water and carbon dioxide. Among the several PD methods, the Z-scheme is a mechanism which is suitable for increasing the efficiency of PD, particularly in the context of heterogeneous photocatalysis. The effect of different parameters demonstrates that the rate of PD of dyes rises with increasing irradiation time, catalysts dosage, and oxidants up to a certain limit. However, beyond that limit, the rate of PD of dyes was found to decrease when the initial dye concentration was increased. At the same time, the process is influenced in two different ways by radical scavengers and the pH of the medium. Moreover, in the comparison of anionic and cationic dyes, the cationic dyes are considered as the most dangerous to the environment and humans as they are carcinogenic and mutagenic.

Employing naturally occurring polymers, particularly ones produced from polysaccharides and their various derivatives, shows promise for the future. These polymers have been put to use in a variety of applications, including drug delivery, water treatment, food processing, and cosmetics. This is due to the fact that biopolymers possess unusual qualities, such as being nontoxic, biodegradable, and renewable, in addition to being easily adaptable throughout the process of creating polymeric nanomaterials. The primary characteristics of nanomaterials based on polysaccharides, such as chitosan, pectin, alginate, and cellulose, as well as their applications in the purification of water as nanocatalysts, are discussed throughout the book. According to a review of the research to-date, biopolymers based on polysaccharides have the potential to be an attractive material for the production of nanocatalysts and nanocomposites. These nanocatalysts and nanocomposites have the capability of achieving the effective removal of a variety of pollutants, including micropollutants, dyes, and heavy metals. Despite the advancements made in the use of biopolymers for water treatment, further study should focus on the extraction of green/natural polymers as well as the development of potentially functional biopolymeric nanomaterials. Synthesis of nanocatalysts and nanocomposites employing natural supports such as montmorillonite and clays could be studied for cost effectiveness. On a large scale, consideration should be given to issues concerning the scalability of production, the optimization of processes, and the energy requirements of biopolymeric nanomaterials. It would be an inexpensive technique that would also minimize solid waste if nanocatalysts and nanocomposites were synthesized using waste feedstock and animal manure. Additionally, it would enable the use of various biowaste nanomaterials as renewable alternative feedstocks for the production of biopolymers and their application in the treatment of wastewater, the scalable fabrication of polysaccharide-based nanomaterials at a relatively lower cost, and the exploitation of natural resources. Improved polysaccharide-based magnetic nanoparticles for water/wastewater treatment are also greener biological procedures to lower the cost of producing biopolymer-based nanomaterials under benign conditions and increase their nanocatalytic capabilities. The manufacture of biopolymer-based nanomaterials using these approaches involves altering animal and agricultural wastes, such as eggshell, bristles, and bone, for use in water remediation. Biopolymer-based magnetic polymeric nanocatalysts should be studied in the future since they are easy to separate during water treatment.

www.ingramcontent.com/pod-product-compliance
Ingram Content Group UK Ltd.
Pitfield, Milton Keynes, MK11 3LW, UK
UKHW051850100926
8556IPUK00005B/555

9 780750 355155